Vintage Feed Sacks

Fabric from the Farm

Susan Miller

Schiffer Publishing Ltd

4880 Lower Valley Road Atglen, pennsylvania 19310

Other Schiffer Books on Related Subjects
Collecting Household Linens, by Frances Johnson
Colorful Vintage Kitchen Towels, by Erin Henderson and Yvonne Barineau
Stripes, by Tina Skinner
Conversational Prints: Decorative Fabrics of the 1950s, by Joy Shih
Chenille: A Collector's Guide, by Judith Ann Greason and Tina Skinner
Quilting Traditions, by Patricia T. Herr

Published by Schiffer Publishing Ltd.
4880 Lower Valley Road
Atglen, PA 19310
Phone: (610) 593-1777; Fax: (610) 593-2002
E-mail: Info@schifferbooks.com

For the largest selection of fine reference books on this and related subjects, please visit our web site at
www.schifferbooks.com
We are always looking for people to write books on new and related subjects. If you have an idea for a
book please contact us at the above address.

This book may be purchased from the publisher.
Include $3.95 for shipping.
Please try your bookstore first.
You may write for a free catalog.

In Europe, Schiffer books are distributed by
Bushwood Books
6 Marksbury Ave.
Kew Gardens
Surrey TW9 4JF England
Phone: 44 (0) 20 8392-8585; Fax: 44 (0) 20 8392-9876
E-mail: info@bushwoodbooks.co.uk
Website: www.bushwoodbooks.co.uk
Free postage in the U.K., Europe; air mail at cost.

Covers and book designed by: Bruce Waters
Type set in Americana XBd BT/Souvenir Lt BT

ISBN: 0-7643-2611-2
Printed in China

Dedications

Well you have to know I have a sense of humor for this first dedication. Just for fun, I would dedicate this book to Oprah Winfrey (who could make this price guide the first price guide to ever be a best seller) and Dolly Parton (who could make my feed sack collection the center of a new permanent small farm and quilting museum area of Dollywood, her amusement park at Pigeon Forge, Tennessee). These two ladies know about country and have lived the "rags to riches" story we all admire and adore. No doubt they also know what a feed sack is. I know Dolly should, as I saw them in her cabin museum and at her Aunt Granny's diner. Oh well, as an old farm girl I can dream about sharing our small farm heritage with the world.

I would make a dedication to my parents, who have passed on. As I grew up, they taught me the value of remembering and loving my childhood and the farm life that we lived. They taught me the importance of hard work, something I have not forgotten even in "retirement." They taught me responsibility, which I hope I never forget. These values hopefully show throughout this book.

I would make a dedication to my son, Robert, who loves feed sacks as much as I do. He shares this passion and selected and cherished this collection with me. He has helped in every way, including buying, selection, computer work, and information gathering. A better son could never be.

A dedication to the men and women farmers who lived through the ages of feed sacks on the farms. To the men who bought the feed, the ladies who washed and saved the sacks, and the farmers who saved the printed feed sacks—we collectors owe all of them much gratitude. I hope men enjoy this book as well as women, for the farm sacks should also be as interesting to them.

A dedication must be given to all the manufacturers and designers of the sacks that we now collect and enjoy.

My biggest desire is that everyone can enjoy this book. Not just for the fabric of the bags, but for the many stories that can be shared about farm life. With the vast amount of feed sacks available, this book is certainly not complete and even if there are more books, probably never will be complete. So a last dedication is to the future of this learning, collecting, and sharing.

Preface

Hello fellow feed sack collectors and those who want to learn about this interesting collectible! Collectibles are most enjoyable when one has a true passion for the subject, so it might help you to know a little bit about me. Of course, writing about oneself is the most difficult part of putting a book together. Having been raised on a small farm and having sewed gathered skirts and matching hair bows from the feed sacks obtained by my Dad when I was approximately thirteen, I guess I have actually lived the memory of these fascinating fabrics. I am now sixty-one years old and our son is thirty-eight years old. My husband and I are both retired from a Fortune 300 printer. He was a head pressman and I was a buyer.

Active collectors for years, we have collected items from many different categories. I also wrote a collectors' book on trolls almost thirty years ago now. Whether dolls, china, pottery, glassware, postcards, or whatever, we enjoy the hunt and the memories these items produce. We have also enjoyed the travel, the people we have met, and the many memories shared. We enjoy the clubs, newsletters, and the history learned. I hope this book will spark memories, fun, and great times for you as well.

All of the items in this book are owned by me and are in my collection. I would love to have shared many more items with you, but for publishing purposes that would just not have been possible. I am sure that those who already collect have many different patterns and items in their collections as well. Please note that I have shown only a small number of collectible items *made* from cloth sacks, as I wanted to keep the book very concentrated on the sacks themselves. I am sure men would want to see more agriculture bags and ladies would want to see more craft items and patterns of fabric printed bags. Everyone would want to hear more stories. I would love to share pictures of my hundreds of quilts, crafted items, and thousands of printed sacks. All form a great part of our history. So if you like this book, please write and request a second book. It would help encourage the publisher and would show the interest and actual need for more books to follow.

Whether you are a collector of fabric, bags, or just enjoy sewing with wonderful textiles, please enjoy these collectibles that come from a much poorer—may it be simpler—time in our history. I owe much to my folks, who are long gone but left me with a great legacy for truth and the joy of farm life. Memories are formed daily and may yours all be joyful.

Contents

Introduction

Congratulations, you have a real learning opportunity in this book. A book in color with many pictures, descriptions, and a price guide on the subject of vintage feed sacks and cloth bags. This book has been needed by collectors for a long time and will be most welcome in many homes—the homes of artists, designers, seamstresses, farmers, millers, people who quilt, and folks who have collected for a long time without any guidance or help with their collections.

It also is an invitation to share stories, historical documentation, and studies on these topics so as to further our knowledge. I hope this book is a help, and maybe just the first of many more to come. Everyone has a story and these stories are most welcome. History is a wonderful thing, and the history of these bags from homes and farms across our nation is a great story to hear and tell.

So here we go, into your home, into your sewing room, into your barn, into your designer studio, and into your life.

Cloth Sacks – Beginning To Current

Cloth sacks supposedly were made to replace expensive wood barrels. I am guessing they were also easier to handle. In time, they also evolved into a second usage or second market. In other words, the sacks were used for the cloth of the sack, not specifically to hold something. Cloth sacks began in the late 1800s, with my favorite, the chicken feed sacks, probably ending sometime in the early 1960s. This is disregarding any reproductions or attempts at remakes. I feel most of the surviving chicken feed sacks that we are collecting were from the latter years, as those from the early years were turned into clothing or quilts simply out of need. In those days, nothing could be hoarded just as a collectible or put away for later use. The only ones saved early on might have been those that were put away for a bride-to-be in a hope chest. Most families, however, had to use these sacks as soon as they were cleaned to clothe growing children or fulfill household needs.

I was twelve or thirteen when I sewed with them, and since I was born in 1945 that would make the time frame approximately 1959 or early 1960s. I am sure that this was probably when mass production of chicken feed bags with fabric designs was ending. I will have to admit we certainly were not a rich family. We were simply middle class, like most small farmers in those years in Indiana. This was the era when a modest farm of two hundred acres produced a comfortable living for a small

family. Today, that certainly would not be true. If we were poor, we did not know it. The cloth sack fabric was not used out of a monetary need, but simply because I loved the prints. This was my first attempt at making clothing and I was so excited and proud of the end product. The gathered skirts and matching hair bows I made were loved by me and I had no shame or regret in wearing them. Such a thought never even entered my mind.

Printed fabric sacks began in the 1920s and were being produced in mass numbers through the '40s. By "mass numbers" I mean many millions each year. Keep this in mind, as there are two possibilities that this fact will bear out. There could have been a large shortage even though they were mass-produced, because they were being used or converted so much. Those wonderful farm women were quite thrifty and were helping their husbands get the quantity of similar prints they needed for sewing projects. They also were helping pick out patterns of print they liked. However, we also know that many feed sacks have been saved and cherished, and are now coming into the marketplace. Many of the generation that saved them are passing on and their belongings end up being passed along to others or sold.

The importance of this fact is that it will affect the market value of these items. Just as with any collectible, availability affects value. If they were to suddenly flood the market, prices would slump and the value of any already purchased might slip. I have seen many collectibles experience drastic price reductions when their popularity suddenly produced a surprise event of them coming out of the woodwork. On the other hand, an increase in collectors could produce an even greater shortage, and the price will then skyrocket. The current value of commercial sewing fabric is growing not reducing. Good quality fabrics have steadily increased in price.

Sometimes what we call a fad or collectible of interest produces new collectors from other countries as well. Online auction site eBay has provided a new means of accessing collectibles and now is not just in the United States, but throughout the world. The eBay site is growing like mad and going to new countries all the time. I do not sell feed sacks, but I do sell other items on eBay. Although I have not yet opened up my sales to anywhere other than the continental United States, I receive e-mails from people in many other countries asking to buy my products. Collectors in other countries who pick up on the appeal of feed sacks will have a means of purchasing them through eBay.

As we also collect Raggedy Ann, we have witnessed the Raggedy Ann collectible market spread across the world. Raggedy Ann items are especially desired by the Japanese and I have heard rumors that feed sacks are catching on in Japan as well. Who can blame them for loving these soft, wonderfully designed fabrics provided by the feed sacks we are collecting.

Types of Cloth Bags For Collecting

There are many types of cloth bags that are being collected. I personally like the chicken feed sacks that were made to look like different patterns of sewing fabric to make clothes etc. However, it would be unfair not to discuss and study the other types of bags that many people are collecting. I have even seen many people already starting to collect printed-paper bags. Let's face it—your collection is yours alone and no one can tell you what to collect. The only important thing is that you are having fun.

So here are some of the cloth bags I know of that people are collecting: bank money bags, vintage shot bags, peanut or nut bags, potato sacks, sugar sacks, salt sacks, rice sacks, seed bags, corn meal bags, flour sacks of all kinds and grinds, and all agriculture and farm products.

In essence, cloth bags were made for anything that a sack could hold. However, due to costs and the problem of vermin and bug access, most of the above types of bags have been replaced by plastics or treated paper.

Chicken Feed Sacks
Fabric Printed Bags

As I will say over and over, this is my favorite type of bag and the basis of my collecting. I want to describe to you the form and aspects of these bags.

They are basically all cotton, made from the cotton grown in the fields of our southern states. Chicken feed sacks consist of what I would call soft and absorbent cotton. Although some may have a percale-like tight flat finish with many threads per inch, the most common collector kinds have fewer threads per inch (without going into the third kind, which might have very few threads per inch and are of very loose fabric that unravels quite easily). While seed sacks are very heavy and you can actually see a visible sort of bumpy texture, chicken feed sacks were soft enough that underwear and every other piece of clothing, bedding, and pillowcases could be made from them.

A selvage edge is a tightly bound edge and the cut or raw edge is the edge that can unravel. I tell you this so you can visualize the construction of a bag. The selvage edge of the bag's fabric will be at the top and the bottom. The cut or raw edges form one side of the bag sewn together. The other side of the bag is simply a fold.

So when you open up a bag and lay the fabric flat, you would have a selvage at the top and the bottom and two raw edges, one at the right and one at the left. The string holes will be on these raw edges and across the top and bottom, with the possibility of a curve stitch where the corners were at the bottom. The string holes sometimes shrink greatly with washing and it can be hard to see them. This can be good and bad as it makes it difficult for you to determine if have a real bag.

The String on Cloth Bags

When I talk about string I am referring to the stitching that forms the bag and closes the top of the bag. I am not referring to the thread that makes up the fabric content. When cloth bags were first made to replace wooden barrels and tin containers, they were sewn by hand and with inferior threads that would not hold up, especially for 100 lb. bags. Once sewing machines were developed that could use a stronger thread, cloth bags could be made with sturdier thread and would not break open.

The string that made or formed the closures on bags was just that—string, not thread. This string was also formed or sewn in a very special way. The chain stitch that was used allowed the farmer or end consumer to easily open the bag, and only when desired.

The string used also helps identify the product. It is identifiable if it is still in bag form and also from the holes that might still be visible after it is removed. Most early string appeared to be all cotton, but I have seen later bags where the string looked shinier and probably had extra fibers or synthetics added for strength.

Just as with fabric, string was saved and used for other purposes by thrifty farmer wives. For example, many doilies and other items were made of string. I have seen several items made from this string and unless you knew the string's origin, you would not dream that it had come from a cloth bag.

Picture showing the chain stitch string inside the bag when the bag is turned inside out. $65-70.

Printed Advertising on the Bag or Sack

Indelible inks were supposedly used on bags where you wanted to retain the advertising and keep the letters crisp and clean. Sometimes water-soluble inks were used on these, however, and the lettering you wished to keep was lost.

Water-soluble inks were used for printing advertising or instruction on bags where you wanted to remove the advertising or words easily. Unfortunately, the lettering was still sometimes very difficult to remove. As you can imagine, children did not like going to school with the word "Chicken" displayed on their clothing and were embarrassed by the words "Feed," "Mash," or "Hog" on their bloomers. There were many recommended methods of removing the printing, and supposedly the method was also printed on the bag. Some of the methods I know of included the use of Fels-Naptha soap, household chlorine bleach, or soaking in paint removers such as kerosene. Old time lye soap and the art of rubbing were also ways to remove the print. The word "easy" is not the word to describe this process.

In later years, manufacturers realized the problem with printed bags and began using paper labels for the advertising, especially on chicken feed bags that were made of printed fabric for sewing. Even this paper was not always easily removed, however, and sometimes it took quite a bit of soaking to get them off.

Remember we are talking about years when washing machines were not automatics with a dozen settings. Early washing machines would have simply been wash boards, either metal or glass with a wooden frame—and that meant many worn out knuckles. Years of washing would have "progressed" to the wringer washer. Ours on the farm was white with red trim on four legs and wheels, and a side hose for emptying the water.

Getting white sacks to be white or cream without print was difficult, but the end result was always rewarding—especially when colorful threads were then used to embroider great pictures and sayings on the sacks destined to be used as tea towels. Many 7-day towel sets can be found on the antique market with cute pictures typical of that 1940s and '50s period. Thus, we have yet another collectible field. Many sacks were preprinted with permanently colored side stripes to be used as pre-decorated tea towels.

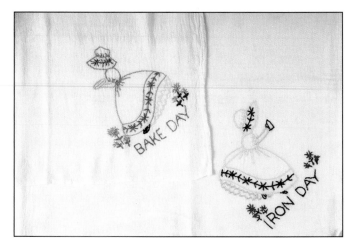

Sample of tea towels for work days of the week with girls doing household tasks. Embroidered on bleached feed sack fabric. Complete set would be seven tea towels representing each day of the week. $80-95 set.

Sample of textile painted tea towels for each day of the week. These are vegetables, painted on bleached feed sack fabric. Complete set would be seven tea towels representing each day of the week. $70-85 set.

Mr. Richard Peek

Mr. Richard Peek, supposedly a salesman for the Percy Kent Feed Company, is said to have come up with the idea for printed feed sacks. I could not verify this fact, but I doubt he got his dues or was paid sufficiently for this wonderful idea that resulted in millions of bags and many happy farm wives! Many happy collectors would also like to thank him today. The Kent Feed Company was certainly a name we knew on the farm. I am very proud of the Kent cookie jars I have collected that are decorated with the printed bags theme.

Some people call printed fabric chicken feed sacks Kenprints. The special Kenprint I own is on a paper banded bag that is a solid red with overall white small polka dots. It is a Critic Feeds, with pictures of all kinds of animals. The bag is for 100 lbs. of Growing Mash and manufactured by Schultz, Baujan & Co., Beardstown, Illinois. Up where the freight information is printed, it reads "Ken-prints. To remove band soak in water." So it verifies this story of "Ken-prints."

There are approximately thirty plus cloth mills that made printed feed sacks. Thousands of elevators and feed stores carried them to the public. One of the mills that made the bags was Bemis and I have seen many bags that read Bemis. I single this out, as when I was a buyer for a printing company one of the categories I had to buy was packaging, and even today Bemis is in business with other types of packaging. They, like all manufacturers and distributors, have changed types of packaging as the industry modernized and changed.

At a country store located in a notable amusement park, I recently saw many dishes with feed sack prints that look like the sacks printed with words. I have also seen fabric, tiny decorations, clothing, t-shirts, and all kinds of items popping up with designs that look like printed words feed sacks. There are at least three current cloth factories making fabric to match or look like feed sack fabric. It takes that eagle eye or that experienced collector to be able to find the real product.

Standardization of Measures

Of course, it was necessary for the government to standardize the measures of products to be sold so that like amounts could be compared in pricing. When this occurred, it naturally controlled the size of the bags needed to sell the standard measures. Around 1943, the government standardized six sizes, ranging from 2 lb. bags to 100 lb. bags. I understand it might have been the War Production Board who standardized the sizes of cloth bags to 2, 5, 10, 25, 50 and 100 lbs.

It is the chicken feed sacks that were done in the 100 lbs measure. For those of us who collect fabric cloth printed chicken feed sacks, that is the most important size. It is the size that controls the amount of fabric we generate from each bag. Even at that, the bags will vary slightly in size. They mostly range 36"-37" x 40"-46" when opened up flat. Or, basically they measure 36"+ from top to bottom and 20" + wide in full bag form. If a bag varies greatly from this size, I would question if it is a real original 100 lb. chicken feed sack.

I can't help but think someone was very disappointed with the 100 lb. chicken feed bag when the lady of the house wanted the design of fabric on the bag that was at the bottom of the heap. I am also sure farmers were lifting most of the bags only by hand, because equipment as we know it today was not available for lifting. But most farmers had very built up muscles from their daily chores.

I can remember fondly my farmer Dad in the bib overalls that he wore all of his farm life (usually with dark green work shirts with long sleeves rolled up), carrying and lifting 5 gallon buckets and sacks of seed or fertilizer, mostly all paper bags in later years. His forehead would be white from the John Deere baseball cap he wore in the sun and he always used big gunnysacks to go to the field and hand shuck the sweet corn we enjoyed in the summer. He would walk the rows and pick the ears at just the right maturity, trying to beat out the raccoons. The raccoons judged the corn's readiness during the night and could beat out we humans most of the time.

As a little girl, I thought there was surely no one stronger than my Dad. He was always a joker as well. For example, when I was older one day he told me the electric fence was off and I grabbed it. No, it did not hurt me but sincerely gave me a kick. He never dreamed I would grab it. When my older sister was young, he also pretended to eat mud pies that she made while secretly tossing them over his shoulder.

A Note About Pricing

Pricing throughout the book is for whole, clean, unblemished full formed or whole opened bags. Pricing is subject to different areas of the country and your own personal desire to own a product. No one can tell you what to pay and these prices are only suggestions from personal knowledge of current pricing across the land. Prices are not meant for appraisal purposes. They may be very high for your area of the country or very low. Remember the important thing is to just have fun and enjoy your collections.

Stories told were for enjoyment and not meant to injure any parties. Stories are a great part of this collecting and I am sure like most oft-repeated tales, they change from telling to telling.

The Farm, The Passion, The Stories

I know this is a collector's guidebook and price guide. But you will not know the entire story if you do not understand the passion. After all, if you too don't feel some of the passion for this specific collectible, why work so hard to collect it? Collecting is hard work—enjoyable, but work! Working on your collection you learn all you can, you look, and you search. Such collecting does use your resources, i.e., your money, your time, your energy, and your thoughts needed to form a fabulous collection. Your passion, joy, and feelings of great fun and memories—old or newly formed—will give your collectible its value. I guess what I am saying is, do not depend on monetary value only. Your collection should mean much more to you than that.

The farm for me was some two hundred acres in Fountain County, Indiana. The farm was near Veedersburg, Mellott, Newtown, and Hillsboro. Dad was born and raised on this farm. His mother had him late in life and she was soon divorced. They stayed on the same farm where Dad was born. In high school, Dad learned to play the saxophone, clarinet, and flute. After high school he traveled in the big bands of the '30s playing these instruments and singing. He met Mom in Detroit and returned to his mother's farm. Mom was all big city and the farm was small and probably, as I was told, poor. I believe the old adage of "you cannot eat the hens or that is the end of the eggs" applies. But Mom surely did adjust to living on the farm, as they had been married for sixty-two years when she passed on.

Part of that adjustment included farm life and raising chickens to clean and sell. Now don't faint, but they were one of the first couples to raise the baby chicks in their house in the upstairs. One must do what one must do. The joke was that when the preacher visited, the small chicks would start running in a circle, causing a very perplexed minister to wonder what was in the upstairs. Soon, I understand, some of the neighbors did the same thing.

There was no greater animal lover than my mother. However, you must kill chickens to prepare them for eating. Years later, after I was born, I can remember Mom wringing the chickens that were to be butchered and sold. Wringing their necks by her swinging hand even though she stood only 4' 11" and was of small stature. No axe was needed by Mom to kill the chicken. It flopped away headless only too soon, to be scalded and plucked. It was Mom who taught me how to professionally cut up a chicken into the "proper" pieces.

So yes, I am talking real farm, real chickens, and real chicken feed in real feed sacks. As a matter of fact, later when they had a brooder house there was a cast iron, wood burning, potbelly stove in the center of the brooder house. When I was first married, Dad gave me this cast iron stove. That cast iron stove has been in my house painted black as a decoration for all forty years of my marriage. The brooder house long ago dissolved to ruin and is now gone.

I have fond memories of when the hens in the hen house would rarely have a soft shell egg in their nest and of the oyster mash you fed the hens. The ground corn in feed sacks and the odors of the feed I still remember too. I remember the joy of growing up so innocently on the farm.

When I first had home economics in the 7th grade, I learned to sew on a treadle sewing machine. I believe at that time Mom and Dad already had an electric machine. Mom patched but really did not sew to make clothes. My grandmother, or Mom's mother, was the real seamstress. After I learned to sew, I made gathered skirts and hair bows to match out of feed sacks and wore them to school with no shame. I loved them. We also had feed sacks hemmed as drying towels for our dishes. They were soft, absorbent, and great. They lasted long after Mom had a dishwasher. I never did like a dishwasher and currently live in a house with a dishwasher that I had my husband disconnect because I like washing dishes. The dish towels Mom had were yellow and orange paisley feed sacks and I remember them fondly yet.

I apologize if I ramble about my childhood and the farm. With Mom and Dad now gone and the farm divided among their three children, I guess all we have are our memories of the happy times. Those times were rather poor and primitive, but nonetheless great times for me. They were the seed for my later collecting of chicken feed sacks. They are why the preferred type of collecting for me is whole chicken feed sacks, even though the farm had many types of agriculture cloth sacks.

Quite frankly, a large part of collecting cloth sacks is listening to and sharing stories. This is a price guide and yet that is the part that I cannot price as it is priceless. Everyone over fifty has a story to tell. Even some younger folks have stories that they repeat from their mothers or grandmothers. So whether you like it or not, get ready to listen. No matter what you are doing, whether buying, selling, searching history, going to museums, or whatever, people will want to tell you a personal story. So get used to it and listen, because these stories will enrich you and teach you a great deal about your collectible. Yes, stories are very much a part of this collectible field and my book would be remiss if it did not repeat a few.

Dad on the Farm

This story about Dad and the farm is for the men, so they may see how a child's memory of Dad affects his or her life. When I grew up on the farm we did not go on vacations as most children do today, as there were daily feedings to be done and one did not normally leave that to a stranger. Interactions were among Mom, Dad, brother and sister. No other children of your own age. I suppose this was a natural situation that made one dream more or have memories more entrenched in family activities.

I remember I had made a scrapbook of John Deere tractor pictures that were retrieved from Dad's *Successful Farmer* magazine after he was done with it. Dad had a B and an A John Deere. Later in life he had a 720. He actually started out farming with horses but that is before I can remember or perhaps before I was born. He was still playing in the big musical bands early on and would do that late into the night. He told me that if he fell asleep when farming during the day, the horses would stop by themselves at the end of the field. Good thing he was not on the John Deeres then.

We had a barn that had been on the farm from early days. Admittedly, we also had a new barn that was built when I was about five or so. When the new barn was done, we had a big neighborhood party. I can remember they loaded the hay and feed trough with home cooked dishes and had a large community feed. I think they danced in the haymow. I really cannot remember this well—I was too young. I do know that later on the trough was filled daily with feed for the fattening of Hereford cattle. Hay making and putting the bales in the barn was done with a large pronged fork on a rope that Dad loaded with bales. A tractor on the other side of the barn then pulled the bales up and across a track and put them in the upstairs or haymow of the barn.

The old barn was made of dark wood, unpainted, and was thought capable of falling down at any time. It had hewn logs as beams and was constructed with square head nails and wood dowels or pegs. I remember sows having their baby pigs in the side area. There was a wall and a small corn crib. In the middle was a driveway where Dad parked the big farm truck. Next, there was a long feed trough and another area for growing pigs. I said I thought the barn was capable of falling down. Well that was certainly wrong. When Dad finally did decide to tear down this barn, even the aid of a tractor could not pull it down. The construction was so good it was unbelievable. Today the wood would have been fought over for country craft projects, but in those days it was quickly burned and not treasured or saved.

The big farm truck I mentioned was used to haul many cloth sacks, feed, or whatever were the needs on the farm. If you can picture the movie *Hoosiers* you can picture that truck. It was even rumored that where Dad had traded the truck in they might have sold it and it might have been one of trucks in that movie. New Richmond, Indiana, is quite close so this is very possible. We think it was a Chevy and about a 1946 model. I do remember it embarrassed Dad once. When he would turn the truck, it often blew the horn on its own. One neighbor lady who was bending over while gardening received that horn honk when Dad turned a corner. At least that was Dad's story.

I can still picture Dad carrying sacks over his shoulder. Looking at him, one could almost feel the pressure from the weight. The cloth sack would wrap down over his shoulder. Well, I just thought I would include this short story about the farm and Dad. I am sure it was Dad having been born right on the farm that is the true heart and beginning of this book.

Chicken Houses to Play Houses to Cat Houses— How the Farm Changed

I cannot leave the subject of the farm and the reason or initiation of my passion for feed sacks without telling you that small farms are basically a thing of the past. As noted earlier, my parents have both passed away. I helped my husband and a neighbor clean up the farm just before my father passed away.

The brooder house where the potbelly stove was housed had long ago fallen in and disappeared. We pulled some of the last floor beams out of the ground and burned them. The hen house had long ago been turned into a tractor shed, with a small garage added on the front that was open like a carport. We gave it a new coat of white paint and picked up limbs from its roof and the surrounding area. On one of these limb-picking-up trips, my husband had a small snake start crawling up his arm. Obviously, that load of limbs was quickly thrown far away from him—including the snake!

One of the chicken houses had at one time been turned into a playhouse for my brother and me. We had our toys in it and a kitchen with a child's cabinet and toy dishes. It was a wonderful playhouse. After we grew up, Mom filled it with straw for bedding and turned it into a cat house. My mother loved cats and they loved her. She kept the kittens and bigger cats in the house when the kittens were little in order to keep wild animals from eating the kittens. One night, a skunk met her in her tasks and indeed did spray her. She was half stripped by the time she hit the house. She fed the kittens and cats there too. Another of the small chicken houses had been turned first into a grease house for the tractor oils and grease, but later it too became a cat house. It had long ago been cleaned up and disappeared completely.

During this clean up, the chicken house that had once been the playhouse was hoisted up by the neighbor's rig and placed on the fire…disappearing in smoke. I know you readers simply have to be laughing by now, but believe me this was the end of an era. It was the final last moment of a time gone by. I cry as I write this paragraph. My Mom already gone and Dad in just a couple of months gone also. The cat houses were now gone and the chicken houses gone too. The homestead portion of the farm was inherited by my brother and just as well, as I don't really want to go back. The "good old days" are the days I care to remember. When chickens were raised, when I sewed with feed sack fabric, when Mom and Dad farmed, and all seemed right with the world.

Learning to Sew

I have to tell you that I am a person with little memory of my childhood. Not that the memories were unimportant or sad, I just did not store many of them. (My son, on the other hand, is a person who remembers every little detail from age three on.) So anyway, the memories that I do have must be important ones and are very dear.

When I was growing up, my Grandma on Dad's side had passed away before I was born. My Grandma on Mom's side lived in Detroit, Michigan. We lived near Lafayette in Indiana on a small farm. Detroit might as well have been a million miles from the farm. It made it seem that we had grandparents only two or maybe three weeks a year when they visited, as the distance made it impossible to be together often. There was a great contrast in lifestyles also. Grandma lived almost in downtown Detroit, residing on a small street called Jane just off of Gratiot near the inner city small planes' airport. We lived on the farm, and just simply did not get off of the farm.

One year when I was a preteen, Grandma took me to Detroit to visit for a week by myself. I can't remember for sure, but my brother might have gone too. Grandma took me one day on a visit to the J. L. Hudson Department Store. At the time, this was one of the largest department stores. We got to see a fashion show. You can imagine what a thrill this was for a farm girl. To top things off, the show included fashions that were sewn from patterns in the store and fabrics in the store. At the completion of the show, Grandma purchased a pattern and the same fabrics worn in the show to make me a jacket and a skirt. I can remember the exact colors and trims she used.

Well, I am sure that the joy and great memories this produced for me are what encouraged me to learn to sew. Yes, it was quite a contrast coming back to the farm and sewing feed sacks. But the contrast was ok and I found as much or more joy doing my own sewing using the farm feed sacks. This was the start of a lifetime of sewing and enjoying fabrics. I even sewed all my son's baby clothes, making little shirts and diaper covers, booties, long gowns, and even his receiving blankets. It was the seed and I am so glad the seed was planted or should I say "sewn."

Remembering the Prints

I don't know about you, but remembering names and other details is becoming very difficult to remember now that I am past sixty. But some memories have indeed stuck with me. When I was in the 7th grade I took home economics at school and with it came my first formalized lessons in sewing. I can remember the prints that I sewed with and the items that I first made. They were made with commercial fabrics. There was a gather on band apron, tan with pink rosebuds. Next came a gathered skirt, black with yellow flowers that I think were sunflowers, and a plain yellow blouse. Next a dress made of tan with tiny orange flowers over all. Each required increasingly more difficult skills. I even got to be in 4-H a couple of years and had my skills at sewing and modeling competitively judged. All these projects were probably sewn on a treadle sewing machine, which, when I think of it, should have been an impossible task. Bless those old fashioned days.

My sewing with feed sacks took place at home. There were lots of gathered skirts and matching hair bows. There was one particular print that I remember distinctly, perhaps because it was so bright. It was green and red with a hex shaped or curved paisley print. The purpose of this story is to relate to you an event that brought a great smile to my face many years later. Although I have traveled extensively throughout my life, I had never found this print in any part of the country. Well, several years ago, I was at a local flea market not seven miles from my old farm home and found a comforter primarily made out of this print. There it was as bright as ever. Some fifty years later and not far from home—that same print. Although comforters are not my favorite items as they are bulky and a little hard to store, I had to have it. Floods of memories came back as I felt and examined that familiar print.

This event demonstrated another fact to me. Given the millions of prints that have been made throughout the years, many have perhaps stayed very regionalized. That is why if you want to find different prints than those you are familiar with in your area you need to travel to different parts of the country or use a tool like eBay to reach out and find different prints.

The Blonde Telephone Stand

I have been an office worker all my life, handling accounts and purchases for a very large factory. I worked for the same business for thirty-five years, starting there directly after Indiana Business College. Prior to that, I took business courses in high school that comprised my major. Long before their deaths, my folks chose me as their power of attorney to take care of their business if they were incapable or to end their affairs when they were to pass on.

I have previously talked of responsibility in this book. It is not always pleasant or desirable. Before they passed away, my folks had a townhouse and a farmhouse that were fully maintained and fully furnished. When Mom passed away, Dad decided that once her passing became known, the farmhouse might become a target of theft. In addition, he no longer drove and would not be spending as much time there.

I knew I never wanted anything of Mom's to be sold at public auction. There was a real need in the family for the furniture from the farmhouse, however, as my brother (having gone through several divorces) could well use these pieces in his home. If you have ever cleaned out a house of your childhood, you probably know the hurt and pain of change.

There was only one piece of furniture that my brother did not need and really could not use. It was the blonde wood telephone stand. It stood in front of a wall where Mom and Dad once had a crank phone. You know the kind—with bells, a wooden box, and a crank to use shorts and longs. With the advent of the dial phone, Mom and Dad used this wooden cabinet as a telephone stand. The blonde telephone stand was certainly not a prized antique. Blonde furniture has really never caught on in the antique business. However, there was a burning desire for me to own this blonde telephone stand.

Unlike teenagers of today, I don't think I had ever used the telephone to correspond with my peers. Farm children in those days (and let's face it, I am old) never got off the farm. At least I didn't. However, there was an important reason why I desired to take home and own the blonde telephone stand. Actually, I already owned a similar treasure that the cabinet held. Only because it was Mom's and from my childhood was it so important to me. For you see, the blonde telephone cabinet with the pull-out chair, drawers in the side, and an open-up top really was my Mom's sewing machine.

Home Grown and Skills That Last

I said that my family was not poor, or that at least we never knew we were. I guess we were farm middle class like so many local farm families. Certainly not like the big corporate farms of today. My husband, on the other hand, was raised with five brothers and sisters and pretty well knew that they were not wealthy. Some of his stories will hurt one's senses and yet there is no shame as many country families were much the same. They were rich in fun and love among the family, but not rich in material things. He has often said if they could not raise it, catch it, or hunt it they probably did not have much.

One of his stories has to do with barefoot summers and, when there were shoes, putting hog rings in the shoes to keep the tops attached to the soles. Christmas has always been hard, as I am sure his childhood Christmas memories were not that happy. He will not revert to going overboard now. I guess too deep of a hurt to change.

But there is one story that is about feed sacks and sewing skills that is very important and interesting for all of us to know. I asked my husband's sisters if they knew what chicken feed sacks were and if they ever used them for clothing. They immediately said yes they remembered them and that their mother was an excellent seamstress who could make clothes that looked just like store bought.

I had never asked his mother about feed sacks. She now resides in a nursing home and most of her speech is only ramblings. In her mind, she is washing clothes, ironing, cleaning, putting things away, cooking, and doing many tasks that are really not happening. Her days are spent only sitting in her wheelchair.

But one day I was talking to her roommate, who belongs to the local Binky Patrol in our community. They make small quilts for civic organizations and hospital facilities. She was telling me that they make quilts all year long but just can't keep up, that the local hospital, police, and firemen need over a thousand quilts each year that they give out on their various duties. The Binky Patrol also makes these quilts for victims of disasters like hurricanes and tornados.

She then went on to tell me that my husband's mother had been tying the quilts and participating in the Binky Patrol. She told me that her craftsmanship was quite good and that she was happy doing it and participating very well with others—showing that her seamstress skills were still intact.

This news amazed me but showed me that certain skills are really not lost and can still bring great joy. Skills once learned do not leave us. What may have been perceived as a hardship of the past does have great rewards.

Collecting Pottery in Ohio

We actually collect many things that we love to hunt for. Among the items we collect is pottery, which we enjoy very much. For many years, we had been going to Zanesville, Ohio, to seek out the numerous types of pottery that were made in that area and also to attend their Pottery Festival. We visit many of the local towns that made the pottery we like to collect, pottery with trade names such as Roseville, Weller, Hull, Robinson Ransbottom, Watt, Ungermach, Alpine, Shawnee, Lepere, and many others.

On one such visit, we stopped at one of the yearly yard sales. As we pulled in, my heart skipped a beat, for on the table was a very familiar sight: a stack of chicken feed sacks right in the middle of the pottery. We soon learned that the feed sacks were from Maryland and that the elderly couple who owned them visited the area at approximately the same time we do and had brought these feed sacks for sale with them. This quickly started a friendship and we not only bought these sacks but also ordered by mail from them several times. Unfortunately, the elderly man passed on soon after and the lady was unable to drive so far to the festival again.

But it is just this type of chance meeting, which often results in purchases and the sharing of stories about the sacks, that makes this collectible so much fun. It is also a lesson on enjoying today, each and every moment of our lives sharing and caring as we never know when our time will be done.

Travel to Iowa

On one of our many trips, we happened into a rare and wonderful situation. It seems that in a small town in Iowa a gentleman held an annual supper get together in a long abandoned feed store. The supper was to bring together and celebrate friends and neighbors of the town and former customers of the feed store. Such a large meal prompted all day cooking by several women.

Now here is the joy of this story. To decorate the old feed store where the supper was to be held, the gentleman displayed on every wall and all over chicken feed sacks. We just happened to be in the area and searched out the town. Even though the event was to be held that night, the ladies who were cooking took pity on some wayward travelers and allowed us to roam the rooms and look at the display—which made both my son and I cry with such wonderful joy. For several hours, we were able to privately observe hundreds of bags, taking in all the prints and trying to remember what we had seen. Discussing the many prints and examining and enjoying this wonderful display we were totally overwhelmed. Two specific items completely delighted us. One was a Mickey Mouse Disney bag, the first one I had ever seen. It was simply breathtaking. The other special display was a very large crocheted doily made from bag string.

We hated to leave, but before we did we thanked the cooks for allowing us to enjoy a moment in our lives that we would never forget. I hope those ladies end up with this book and know what a wonderful once-in-a-lifetime experience they created for us.

State Fair in Indiana

We have a wonderful State Fair in Indiana. The state's 4-H program is very admired for its ability to provide children and young adults with opportunities to learn many skills and to participate with their peers in a wonderful program. The fair is certainly not confined to programs for young children, however, as there are many adult programs as well.

The fair allows many city folk to see farm and country life that they might otherwise not have an opportunity to witness. There is something for everyone—animals, parades, rodeos, bands, stars, every life skill you can think of, corporate sponsors, farm implements both new and antique, trailers, hot tubs, buildings, yard equipment, pools, advertising and exhibits, and lots of new things to learn and see. Colleges advertise and museums beckon.

But there is one area at the fair that is probably my all time favorite. It is the pioneer village, with country singing, candle making, wood carving, old time implements, weaving, basket making and quilting. An old time pioneer kitchen serves meals to the participants and exhibitors in the village. It is equipped and furnished with everything from the 1930s, '40s and '50s.

I go almost every year. However, one year I missed the fair but my son had an opportunity to go. Just my luck, he got to see an enormous feed sack display in the pioneer village. Now, almost every year there are some feed sacks in the front entry where the quilting ladies have their setup. They quilt during the fair hours. They enjoy talking to the visitors and explaining what they are doing. The year that I missed, there was an extra large display and center of attraction with a feed sack collection. I had hoped that they would have it again the following year. Unfortunately it was not repeated but they did have many pictures of the previous year's display and I got to examine those.

I bet I don't miss the fair again. The pioneer village is always a great experience and education for me, even though I was raised on the farm. It brings back great memories and seems to be one of the friendliest activities. Thanks to all who make the pioneer village possible.

Bridgeton Mill and Bridge

Our community experienced a tragic event recently. An historic covered bridge in Bridgeton, Indiana, was burned by an arsonist and completely destroyed. This was a heartbreaking event for all of us who love history and the covered bridges of Park County, Indiana. Many of the small towns near these bridges also have a gristmill located within them. These use the power of the streams to grind flours and various grains into meal.

My husband and I often visit the bridges even when there is no special event such as a festival going on. My love of feed sacks, flour sacks, and meal bags also draws me to the mills. We visited the mill near the Bridgeton Bridge location soon after the bridge was destroyed. While there, I noticed that the mill has a framed display of chicken feed sacks.

The miller told me that a local lady had created the display for him. He went on to tell me a story she had conveyed to him. Years ago, when chicken feed was sold in sacks with printed patterns, a woman's husband fell in love with a particular pattern. Therefore, he proudly purchased all his feed in that same pattern—instead of selecting two or three bags each of different patterns, he presented his wife with all the same pattern of bags. She did her best with them, but that meant that all the bedding, sheets, pillowcases, all the kid's clothes, all her clothes and her husband's shirts, even underwear, all ended up matching. When they went to church they made quite a sight.

To further the situation, because the cotton bags wore so well, the same pattern lasted for a long time and the family was fairly sick of the pattern before the items either wore out or were converted into something else. We all had a good laugh at this story. The community is planning to try to rebuild their covered bridge. Let's all pray they are able to do so.

We also visit the Mansfield Bridge and Mansfield gristmill quite often. It has been maintained as a state museum and is in wonderful condition with a great history of grinding available. It has a dam and slight waterfall and is a beautiful area to visit.

We have visited many gristmills just because of their connection to the use of cloth bags. The mill at Pigeon Forge, Tennessee, is one of my favorites. Every mill has printed bags designed to hold their ground products, printed with their name and other products to advertise their mill as well. I always try to get a mill sack and hopefully a T-shirt when I visit.

Mansfield Roller Mill Corn Meal Sack. "Grinding with waterpower since 1820." Mansfield Village, Park County, Indiana. 2-1/2 lb. sack. $5-10.

Stone Burr Ground Corn Meal. Picture of mill erected 1817. Spring Mill State Park, Mitchell, Ind. $10-15.

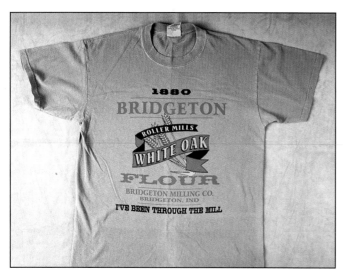

T-shirt from Bridgeton Roller Mill made to look like flour sack. 1880 Bridgeton Roller Mills White Oak Flour. Bridgeton Milling Co., Bridgeton Milling Co., Bridgeton, Ind. As mementos of your mill visit, cute souvenirs are available, including post cards, cookbooks, and other items. T-shirt price $25-30.

T-shirt from the Old Mill Pigeon Forge, Tennessee. This is still a working mill. We thought at one time we would retire to the Gatlinburg and Pigeon Forge area. Tours are available and a fabulous restaurant is located next door. $25-30.

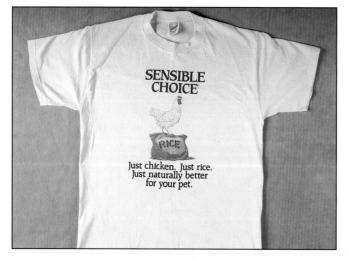

Great T-shirt advertisement, likely for dog food. Looks like feed sack with a chicken and rice bag pictured. $25-30.

Cute T-shirt made by Red Ridge Mountain Outfitters. Shirt looks like 100 lbs. Net Trail Driver Sweet Feed Manufactured by Martin-Lane Co, Vernon, Texas. $25-30.

The Most Fabric

Recently I had a T-shirt that stated: "The Winner is the One Who Dies With The Most Fabric." Ok, it seemed too close to reality to be all that funny. I guess there are worse habits. I do feel that a feed sack collection can and does have value, and value that can grow for those who make wise purchases. I buy both regular fabric and feed sack fabric. I guess if my maker comes soon I might be a winner! At least I have a sense of humor and lots of fun while I am here. I hope we all can be winners.

Chicken Humor

This book would not be complete without some chicken humor. Just yesterday we went to our bank, which is some twenty-five miles away. We then stopped in a nearby small town and ate at a combined filling station, sandwich franchise, general store, and chicken franchise. I ordered a chicken plank with no bone, a potato, two egg rolls and four chicken wings. The teenage girl did just fine until she got to the chicken wings. She looked at the pile of fried chicken and was totally perplexed. Giving up, she finally looked at my husband and me and said, "I don't know what a wing looks like." Keeping our composure, we pointed out the pieces of chicken that were wings.

When we sat down, we both broke out laughing about the situation. We could not believe that a teenage girl would not know which pieces were the wings. These were the full wings with the three sections, not just little drumstick type pieces. My first reaction was she could have "winged" it. That is a pun. Then we mused at our honesty, as we could have pointed out the expensive breast pieces and she would not have known the difference. Who says that truth is not stranger than fiction. I guess teenagers today are different than we were, growing up on the farm!

Use Every Speck of Fabric

On one of our family vacations, we went to Iowa to a Hull Pottery Collectors Club meeting in Amana. Hull Pottery is another one of our collections. Amana, Iowa, is a wonderful place to visit, as it seems to sit in the heart of Iowa and has fabulous country surroundings with great antique malls to visit, scrumptious German feasts everywhere, and great ethnic history to learn. On one of our trips to the antique malls searching for collectibles to add to our collection, we found ourselves looking at quite an array of quilts, comforters, and feed sacks. It was easy to see that if the price were right, many of the whole feed sacks would be going home to Indiana with us. Next we had to decide if the quilts or comforters were made out of feed sacks and if their price, which usually ran too high for me even though comforters usually run much less than quilts, was plausible for purchase. Condition means a lot, because even though the quilt or comforter may still be in one piece the little feed sack pieces may have deteriorated or succumbed to use. New quilts or quilts with no feed sack fabric are obviously of no interest to me and could be put aside immediately. The remaining lot was to be studied. That is when my son found a ragged edged quilt/comforter that was fairly well still intact with feed sack fabric on top—but the real jewel was to be found on the bottom or the back. He discovered an entire life's history on the back. It was obvious that the maker of the quilt had saved every scrap of clothing and every piece of fabric she had ever owned and painstakingly sewn them together, perhaps overlapping or doing whatever it took to join them together due to the various shapes and sizes. It was indeed a history of the family, their buying ability, their need for comfort and warmth. To some, this might have been the worst piece of trash in the world, but my son could see beyond the obvious inferiority of what one normally would not consider collectible to a life that must be preserved and cherished. This was the most valuable quilt/comforter to him and one that would certainly come home to Indiana. Included in the fabrics were many feed sacks and fabrics—only lord knows why she had to save them. The need had been great to provide warmth for her family. Back in Indiana, this treasure will not be parted with and the history is often thought of and remembered.

Variations of this story are repeated constantly with regard to the use of sacks. They were never to be tossed and every scrap used. It shames me when I think of what a wasteful nation we are today. How we discard scraps of material where we straightened the fabric, cut out a pattern, or if it was just too short of a scrap to save. That was not done years ago—instead, every scrap was used. Whether used to quilt, as strips for rag rug making, or for the teeniest of doll clothes, nothing was tossed or wasted. If it was too rough or not pretty it still had a use and that use was found.

So the next time you find something ragged, not what you consider perfect, remember our history and lessons learned. You may be looking at a real treasure.

Learning More About Feed Sacks

If you want to get information about feed sacks, the best approach is to go where history is cherished. Many museums have included information about feed sacks in their exhibits. For example, I know for a fact that the Indiana State Museum has a chicken feed sack that is a white bag with green bows displayed with information about feed sacks from the 1930s.

Current grain elevators or feed stores my have information, especially if the employees have some age on them. Or they may lead you to the sources of bags from current or past records.

Antique water turned gristmills, especially if you can find some that are still in operation or have been turned into historic museums, will have information about flour and meal sacks.

The National Cotton Council or the Department of Agriculture may have some information.

The best information will probably come from Farm Pioneer Groups or quilting clubs and the women who have studied fabrics.

Hopefully this book will be a start of information for you. However, I am sure there is much more to find and learn.

Manufacturer Specific Historical Information

Many of the companies that made feed sacks are long gone. However, there are a few viable feed manufacturers who are still very much in business. I am sure they have wonderful historical information that would be of great interest to collectors, including facts of manufacture and dates of production. This information would probably set us straight on many theories and concepts that we now hold. If there is to be a second book produced, I would want to add this type of information.

However, I wanted this first book to be an introduction to a very enjoyable collectible and to provide just a foothold on information, with more to come later. You won't believe this but I have had many collections of different items throughout my life and once all the history was known and every item was collected the collection was no longer of interest to me. With respect to collectible feed sacks, however, I don't think I will ever begin to have every one, will never have all the information and history pinned down, and thus will never lose interest. Having come from the farm you can see how personal this collection is to me.

So at this point innocent information may be what I have and informed information is what we together might learn in the future. Informed information from collectors as well as from manufacturers both active and retired will be welcomed and passed on when the time comes. In the meantime, let's enjoy our collections for what they are and share our stories of joy and some sadness of times gone by. Let's hold on to our memories, new and old, of the farms, the farm life, and the way of life many of us were privileged to live.

Collector Tips and Hints

The form that your personal collecting will take should be determined before you get too carried away with obtaining items. I suggest making this determination because there are so many directions you could go. Let's face it, you only have so much space and so much time for a collectible. Of course, if you have lots of space that is fine! Perhaps you want a little of everything. Here are some categories that I know of for this type of fabric collecting:

Whole bags
Opened whole bags
6" squares
1" squares
Quilt blocks
Quilt tops
Quilts
Comforters
Aprons
Bonnets
Dresses or clothing
Pillowcases
Tea towels
Doilies or dresser doilies
Decorator items
Various other measures

My personal preferences are whole bags, opened whole bags, and quilt tops.

Primitive ruffled bonnet, elastic in back at base. Laced framed roses bag print. I have many styles in my collection. $15-20.

Craft pieced and turkey stitched small cloth using feed sack fabric colored and plain. Folding table size. $25-50.

Condition

Only you can determine what is allowable condition for your collection. If you find a very rare bag, common sense tells me that you may want that bag if the price is considered reasonable for that condition. However, only you can decide that for yourself.

Common flaws are as follows: picks or holes, excessive raveling, stains, yellowing, mold or mildew, rust, odd cuts, discoloration or fading.

Crossover Effect

What do I mean by crossover effect? Well, you know what I mean when a singer crosses over from Christian to Pop Rock music. Feed sacks or cloth sacks and bags are collectibles that cross over many types of collecting. Maybe it is the advertising relating to some company you worked for or someone in your family worked for. Maybe you are a farmer relating to some of the agriculture products. Maybe you are a banker who collects bank bags. Maybe you are an avid hunter or gun collector who can relate to shot bags. Maybe you are a cook who collects food bags such as flour, sugar, or salt. Maybe you like vintage fabrics for decorating or quilting. Maybe it is the history that appeals to you. Maybe it is the printing or graphic arts that appeal to you. Maybe it is the challenge of the hunt for your collectible. Maybe it just relates to something in your youth that you wish to relive. There is definitely something that triggered your desire or your passion for this collectible. By the way, age and gender are not factors—everyone may choose to pursue this collectible. There are so many different reasons why this particular collection has caught our attention.

Bird Flu Does Not Concern Your Collection

As I write this in April 2006 we are all in the dark as to the future of bird flu. Will it come to the USA? Well it might, but it does not concern or affect your feed sack collection. I only mention it as it might be a worry some of you have. Let me assure you your collectible is safe.

Little or no cloth sacks are used on the farm that would make it to the collectible market if bird flu were here. The feed for animals on the farm is so biochemically controlled I think it may be more safe than the food you eat. I guess I should not joke, as this flu may come to the USA and the result to the farmer may be very bad. It will be even more of a concern if it mutates and can be passed from person to person.

However, I wanted to state that it does not effect your collectible in any way. So enjoy those sacks that you have collected and pray with me that the bird flu will not affect anyone or any animals in our great land.

Monetary Value

Pricing is really just your business, however, if you are buying and selling you must be extremely wise and not overpay. There is no collectible (other than money itself) that has an actual value that can be determined with certainty. At least we hope money has an actual value. Anyway do not depend on your collection as a monetary investment unless you can afford to. That is why I keep mentioning the challenge of collecting and the fun. The collectibles market is very fickle and what has value today may bottom out without much notice. Time, age, and general wear and tear make you think that these bags should have value. Cloth bags, like chicken feed sacks, were produced in multiple millions. Availability always affects price. Common sense tells me there is value in cloth bags, but it also tells me that availability and interest can change. Current market value can be different all across the United States and in other countries.

Reproduction

Many of the fabric companies currently on the market have what we consider chicken feed bag prints and types of fabric thread count wise in the stores. If you are a true collector of the original chicken feed bags, I doubt you are interested in these. What you might be interested in are new fabrics that look like printed pictures of sacks or look like miniature printed sacks or even farm and chicken prints. They make cute go-withs for your collection, for decorating, or for making clothing, but are of no real interest for your true collection of bags. A true collector can identify a real vintage bag.

As far as mass reproduction of the bags or even bags being used for the original purposes, I think you can forget it. The agriculture market today is very controlled regarding feed for animals. Prevention of disease and protection from vermin or insects precludes the use of cloth bags. Medicated feeds can't touch each other and costs do not dictate the use of cloth bags.

I am going to insert a note here about buying things sight unseen, like possibly through the mail. If you are a stickler about having really complete feed sacks (not pillowcases, tablecloths, tea towels, or curtains that are not nor never were feed sacks), watch the size, the grain, the little stitching holes, the two selvage edges. Many people collect 1" squares or 6" squares and may have already "robbed" their portion, then try to sell the item as an advertised complete bag. Ask questions and find dealers you trust.

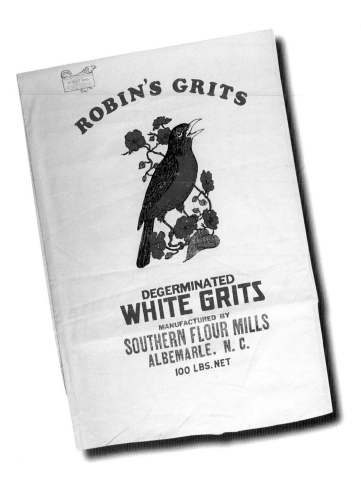

Unusual find, printed yardage that contains about six bags in continuous form. I believe this was to be a real bag for real food. It looks like it came from the bag factory. It was not completed in the bag making process. Robin's Grits Degerminated White Grits Manufactured by Southern Flour Mills, Albemarle, N. C. 100 lbs. Net. Up in the corner in green, printed on a picture of a chicken: Freight Shipping Bag Manufactured by Stwart Bag meeting requirements of Cons Freight Classification for gain products. $150-200.

There is both replicated fabric and clothing made to look like feed sacks. Much clothing was made out of feed sacks for both joy and real necessity in times past. The modern day replication clothing of today, however, is simply made for style or a look of the past. As you know, most styles or looks tend to recycle. Styles and patterns that were once gone seem to come around again.

I am currently involved in a study project that involves modern clothing. I have T-shirts that display mills, look like feed sacks, and advertise every quilt show I have been able to attend. I love to include all of these T-shirts in my collection and enjoy wearing them to shows and at other times. We have great fun meeting new people who notice the topic shown on the clothing and share information about their collections. Great conversations are prompted by what we wear and folks who are strangers soon become good friends.

Also, many modern kitchen items are made to look like printed feed sacks.

Great new tea towel made to replicate a feed sack. 25 lbs. Four Bells Brand Laying Mash. Made by Four Bells Milling Co., Heavener, Oklahoma. I wonder if there is a real city and mill like this. $10-15.

Brand new pot holder made to look like a feed sack. Four Bells Brand Laying Mash. $2-3.

Quilts, Quilt Blocks, and Comforters

I am not a "quilter" as I have never sewn quilts by hand. I have never pieced a quilt with the fabulous names and patterns. However, I have many quilt magazines and books and have studied them a lot. I very much appreciate quilting and it is a wonderful creative "art."

My husband and son would laugh when I say I am not a "quilter" as they know that I have in fact made thousands of baby quilts sewn by machine throughout my married lifetime. I always used 6" squares sewn together in various patterns. I would place seven squares across and seven squares down and would form about a 3" binding front and back. I would use quilt liner purchased in rolls and could get four quilts out of one roll. I usually use a solid white cotton back or a light solid color. I would purchase fabric every time I went to town.

I even appliquéd many designs and sometimes even painted squares with fabric paints in designs or names that were appropriate for the theme and person I was quilting for.

This is the way that I learned a lot about fabrics. But never did I use feed sacks, as those fabrics I wished to save and collect. I cherished them too much to cut up or to give away. For you see, I was giving away all the quilts that I made. The factory I worked for always seemed to have employees who were having babies. All my personal friends and family received baby or lap quilts for every holiday, every birthday, or just for being friends. Not one but sometimes ten at a time. They actually were very useful and just the right size. The thing that is really rewarding is that all those babies grew up and guess what—they still have the quilts and the quilts actually held up through all those years. They used them for babies, grandbabies, decorating, etc.

One friend told me she uses them for lap quilts and decorating and her husband told me last time I saw them that she would surely be buried with my quilts. Her children are now in their late thirties so she certainly has kept these quilts for a long time.

I used to count how many I made in a year and would sometimes take pictures of them before giving them away. Taking pictures of your sewing, your quilts, bags, etc. is a very good idea for enriching your personal collection.

Obviously we are talking about quilts because so many were made from chicken feed sacks. You can recognize that material a hundred miles away. This is one way to find hundreds of patterns of fabric all in one place. In addition, sometimes whole sacks were used as the backing. Prices can range from low to extremely high. You may even consider ragged quilts and what are called "cutter" quilts, which means you will probably cut them up to make toys or crafts or for some use other than a cover for your bed.

There is a definite difference between pieced quilts and what I consider comforters. Comforters are usually older, cruder, and thicker. They can be lumpy from many washings, as the stuffing has lumped to the corners or moved. They may be hard to collect as they take up a lot of room. Many were made with chicken feed sack pieces so you must determine if they are interesting enough to collect or if they fill your intended use. If I did not buy some comforters, I would not have my Uncle Sam designed feed sack print. It is pieced on the front and also consists of the entire back of a comforter that I own.

What I find best for my collecting are quilt tops. That is just what it means. It is the unbound top of a quilt with no filler or backing. I collect either hand stitched or machine stitched tops; when they are made entirely out of feed sack they are an excellent way to find many pieces of feed sack and take up much less room. They may be in better condition and retain all the brilliant fabulous colors you love, as they usually were never used as covers and washed little or not at all. There are many of these around, as the quilting finishing process was the hard process and many pieced quilt tops never reached that state of production.

You may also want to collect quilt blocks. I have had many quilt blocks that never reached the quilting stage. These can be in squares or the many intricate patterns that are available. Lots of time you find them with some finished blocks in a bag and some extra scraps that never got to the piecing or quilting process.

What is even more amazing to me, I am now seeing on the market many quilt blocks that actually were quilted and the quilt ended up so ragged or worn that the person cut the blocks apart. They sell and save the good blocks, even though they at one time had been in a whole quilt. In other words, they cut up a "cutter" quilt without even making something out of it, but just sell the layered blocks that had once been a whole quilt.

I go to many quilt shows and believe me, the quilt I vote for in the contest is always a feed sack quilt. Many dealers also carry supplies for quilts and I often find a supply of sacks as well. While there, I like to get people to talk to me and tell me their stories. If they don't carry sacks they often know where you can find some. And this puts you in hot pursuit of more sacks.

Aprons Made of Feed Sack Fabric

I have often thought some force was guiding me on great adventures. Of course, it could also just be called "being in the right place at the right time." Indianapolis, the state capitol of Indiana, has been experiencing a great evolution of rebuilding in the downtown area. There is a new Conseco building for the Indiana Pacers NBA and Fever WNBA, Indiana ice hockey, and musical events. They built a great new zoo, an atrium of plants and butterflies, a baseball field for a great farm team, a new NCAA Building, an Indian Museum, a new State museum, and an IMAX theater. The White River area also includes pro tennis and there are fabulous swimming facilities in Indianapolis. There is a new shopping center along with many hotels and restaurants that are needed to support all the activities. The area also has a world famous Children's Museum, new additions to the Art Museum, and a newly built Historical Society

Museum. We are so lucky to be less than an hour from all of these great opportunities.

When I first visited the Historical Society Museum, I was shocked to learn that on that day there was a traveling show on aprons. I had not even heard of such a thing. Can you believe I would have such fabulous luck in finding this unexpected event? This form of antique art is certainly well into feed sack printed fabrics usage. You can imagine my joy in examining all the aprons, especially those made from feed sack fabric. Learning the history of the aprons and feed sack fabrics is a very important aspect of this collectible.

I'll have to admit I don't think many modern cooks use aprons. Mainly because many young ladies not only don't use aprons, they simply don't cook at all. Those of us who do cook have too many easy stain removing detergents and simply don't use aprons. What a shame that tying on an apron to do our household duties seems to be a lost habit. The chicken feed fabrics and the many novelty and floral prints were certainly perfect for this type of wear. All vintage linens are becoming collectible so don't forget aprons. Even if a few stitches are missing, be a jewel and restore them. They will be a great reward to those of us that are trying to preserve feed sack prints.

Green and gold hearts and flowers feed sack fabric bib apron. $20-30.

Vintage apron made from very pretty lavender on white floral feed sack. $15-25.

White, green, and purple vintage apron. Large and decorated with purple rickrack. Great design. I love old time rickrack. $15-25.

Vintage apron made from feed sack fabric. This floral is lavender flowers on purple and yellow plaid. $15-25.

Bank Cloth Bags

Before vinyl or leather pouches were available, banks and merchants used cloth bags to transport their money to and from the bank. Many banks provided these bags to the merchants with the bank name, advertising, and bank address printed on the bags.

Many collectors include these in their coin collections. Most are of interest because the local bank was an important memory of a person's business life. Just think of how many banks there once were, and if anything like today they probably were changing names often. Banks are always combining.

Bank cloth bags are simply a service item but collectible today as the cloth bag is probably no longer used for this purpose. Many people collect such bags from their own town or surrounding towns.

S with picture of lighthouse. Seaboard Citizens National Bank, Norfolk, Va. Serving Tidewater Virginia No. 4, 8" x 14-1/2". Bag companies must have numbered their bags to designate bag sizes. $15-20.

Bank Bag. Gulf National Bank, Lake Charles, La. Federal Deposit Insurance Corporation Member. $10-15.

The First National Bank, Waynesboro, Va. $8-10.

American Fletcher National Bank and Trust Co. Indianapolis. $10-15.

U. S. Mint Phila. Cents $50 1975. $5-10

Shot Bags Collected By Gun Collectors

More of a man's collectible and often included in hunting and gun collections are these cloth printed bags for the holding of early shot. Many may have gun advertising and name brand shot. They are especially collected if the advertising includes familiar gun names or pictures of game that was being hunted.

These are normally very small and uniform in size and easy to store or frame for display.

Bank of America. $3-5.

Bank of Virginia, Member FDIC. $5-8.

AA Hard Target Shot Winchester American Standard Super-Shot Olin Winchester Group. $10-15.

Lawrence Brand 25 lbs. No. 7-1/2 Magnum Shot Taracorp Industries. $5-10.

Alcan Company Incorporated Alton, Illinois American Standard Chilled Shot 25 lbs. $5-10.

Feed Sacks by Category

To display the feed sacks here, I have grouped them into what I believe to be different categories. However, what I think is a novelty print may not be what you think is a novelty print. What I feel is a floral may be what you feel is a form of novelty. What I call a geometric design you may call something else. This is a first book, and as our study of this subject advances we may want to change our ideas.

Burlap or Gunny Sacks Animal or People Food

Burlap sacks could be a story on their own. I love to use them in the bottom of my potato bin. They were used for both animal food or people food, as long as they would not sift out of the bag. Peanuts are a prime example. Great for decorating once emptied.

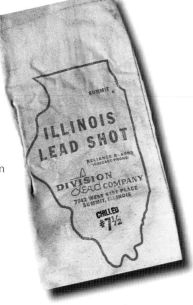

Lawrence Brand "The Shot of Champions" 25 lbs. No. 8 Magnum Shot Taracorp Industries Corp. $10-15.

Illinois Lead Shot Division Oxead Company Chilled #7-1/2. $5-10.

Burlap Bag Jack Rabbit, Navy Beans Packed by Michigan Bean Div., Saginaw, Mich. Valued because of picture. $25-30.

Planters Salted In-Shell Roasted Peanuts. Burlap bag. Current and everyone saves and decorates with them. $5-8.

Human Corn Meal

When we discussed the many mills around the country we discussed their ability to grind corn. They did this to various levels of fineness.

50 lb. Yellow Corn Meal Enriched Degermed Fortified with Calcium. "Donated by United States of America." Freight shipping bag meeting requirements of Consolidated Freight Classification for Grain Products by Bag Service, Inc. Decatur, Indiana. Back of bag has thirteen squares of foreign languages. I am guessing they were shipped to many different countries. $45-50.

Human Consumables

There are many foods that we humans consume that were put up in cloth bags. That group follows.

Potato Bags

You don't normally see too many cloth potato bags, but I have included a picture of a Santa Claus Potato bag that I love.

U. S. No. 1 Santa Claus Potatoes Packed by Market Pre Pak, Inc. Oklahoma City, Oklahoma. Net Wt. 100 lbs. $45-55.

Red Head Enriched White Corn Meal Shreveport Grain & Elev. Co. Shreveport, La. 25 lbs. Net Wt. $25-35.

Sugar Sacks

There were so many companies that made sugar for our consumption and, believe it or not, for many other purposes. Did you know some sugar is used for adhesives?

Franklin Cane Sugar, Manufactured by Franklin Sugar Refining Company, Philadelphia. Listed on the back are types of sugar produced: Dainty Lumps, Pressed Tablets, Superfine Powdered, Confectioners, Old Fashioned Brown, Yellow and extra fine Granulated. $25-30.

Two bags in same picture. Sea Island Sugar Pure Cane Granulated Western Sugar Refinery, San Francisco, California, 2 lbs. Net. $5-10.

G W Granulated Sugar Great Western Sugar Co, Denver, Colo. 10 lbs. Net Sugar. $10-15.

C and H Sugar Pure Cane Berry Granulated 10 lbs. Instructions for opening bag on bag. Copyright 1939. $15-20.

Sunny Cane extra fine Granulated Sugar. Manufactured by the W. J. McCahan Sugar Ref. & Mol.
Co. Philadelphia, Pa. $15-20.

Granulated "The Seal of Purity" Atlantic Sugar
Refineries Limited Montreal Saint John. $25-30.

Domino Cane Sugar American Sugar Refining Company, N.Y., N.Y. $5-7.50.

Jack Frost 100% Pure Cane Fine Granulated
Sugar National Sugar Refining Co. of N. J. New
York, N. Y., Bemis Bags Brooklyn. $35-50

10 – 10's C & H Menu Sugar Pure Cane Berry Granulated California & Hawaiian Sugar Refining Corp 'n San Francisco, Calif. 100 lbs. Net. NRA Member Eagle Picture. We Do Our Part. $25-35.

Flour Sacks

Can you imagine the sacks of flour that a commercial bakery uses? Can you imagine how much flour and other food products we have supplied to other countries around the world?

Original Paper Label on Robin Hood All Purpose Flour. Note that it is a Bemis bag. Bag is blue, red, and green stripe. 25 lbs. $40-55.

White Satin Sugar, 100 lbs. Net Weight, The Amalgamated Sugar Company, Ogden, Utah, Idaho Factories Twin Falls, Burley and Rupert. Striped sided bag. $30-40.

Money Back Flour, The C. Callahan Co, La Fayette, Ind. Neah-Bags Chicago 10 lbs. $15-20

30

Self Action Self-Rising Flour, Cadick Milling Co., Grand View, Ind. 2 lbs. It's White $5-7.50.

A small reproduction "Zip" Flour Fancy Springs Bakers Patent. Has to be a reproduction of a large bag, as it reads 100 lbs. Net, and it is a tiny bag. $5-7.50.

Reproduction High Gluten Flour R. S. Distributors Blue Ridge. It says 100 lbs. Net. Has to be a reproduction as it is a tiny sack. $5-7.50.

American Beauty. It has a red rose picture. Pure Wheat Bran Standard – Tilton Division Russell-Miller Milling Co. Alton, Ill. Use American Beauty Flour. Bag is Bemis-St. Louis. "This bag is printed with Bemis Washout Inks. To remove the ink soak overnight in cold soap suds and then wash thoroughly in warm soap suds. If any ink remains, boil ten minutes in soap suds." Thankfully, they did not wash this beautiful bag or the history would have been lost. $35-45.

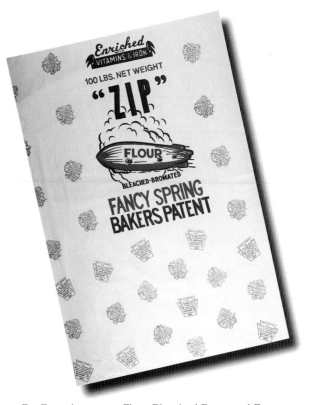

Belle of Springfield Supreme Quality Flour. The Springfield Flour Mills, Springfield, Mo. 100 lbs. Instructions printed at top of bag read: "Open here – handle carefully – avoid waste. Conserve bag for other use." Printed over chicken picture Werthan Nashville Freight Shipping Bag Mailing requirements of Cons freight classification for grain products. $35-45.

Zip Zeppelin picture Flour Bleached Bromated Fancy Spring Bakers Patent. Indian pictures with freight and bag info. Abbot Bag Co, Inc. N. Y. and shield pictures. Keystart Co., Bag Burlap Co, Ohio $50-65.

M Isis Flour for Export 100 lbs. Net wt., Milled in USA by Conagra, Inc. February 1975 Saudi Arabia. $50-65.

Suretest Made From Selected Soft Wheat Bleached Self-Rising Flour. Manufactured by Southern Flour Mills, Inc. Albemarel, N. C. 100 lbs. Freight shipping information printed on an eagle picture. Fulton Bag & Cotton Mills, Atlanta, Georgia. $40-50.

Western Queen Flour, Pure-White Wholesome. Picture of a little girl's face with sausage curls and large hair bow. Registered in United States Patent office. Distributed by Western Flour Mills, Davenport, Iowa. $50-65.

Igleheart Bros., Inc., Swans Down Trade Mark. Swan Picture. Bleached, 98 lbs. Net. This style package put up for bakers only. Manufactured by Igleheart & Brothers, Incorporated, Evansville, Ind. USA $50-75.

Fairchilds Trade Mark Bull Dog Shot Patent Spring Wheat Flour, The Fairchild Milling Co., Cleveland, Ohio. Bag made by Ames Bag Machines Co., Cleveland, Ohio. $55-60.

Traveler Flour. Picture of vintage car. The Goerz Flour Milling Co. $45-55.

Cinderella Flour. Fit for a prince. Bleached, Distributed by International Milling Co., Minneapolis, Minnesota. $50-60.

Crystal White. Pictures wheat so I presume this is a flour sack. Clinton Milling Co., Bowling Green Station, New York, New York, Jeddah February 1982. $20-30.

Flour, Wheat Type A Tip Top Brand, December 1938 Nebraska Consolidated Mills Co. Omaha, Nebr. Unbleached. 98 lbs. Net. Bag and freight information is printed on a feed sack picture with what looks like a cat at the end of the bag (writing is too light to be read). $40-50

Hong Kong Flour Mills Cherries Brand. Extra Special Flour Milled from American Wheat. $50-60.

Robusta Mfd. By Kansas Flour Mills Company, Kansas City, Missouri U.S.A. for export 50 lbs. Printed on a picture of a feed sack. Freight Shipping Bag Meeting Requirements of Consolidated Freight Classification for Grain Products Guaranteed by Fulton Bag & Cotton Mills. $40-45.

Juanita Enriched Flour Paper Label on red bag with Blue Roses. 10 lbs. Net. Great Spanish design on the label. $25-30.

Salt Sacks

There are so many uses for salt—to eat, to cure meat, to soften water, for baking, and the list goes on.

Knowles Brand Kiln Dried Sifted Butter Salt. Packed only for Montgomery Ward & Co. 28 lbs. Net Weight. $20-25.

Diamond Crystal Flake Salt. Sold by Diamond Crystal Salt Co., Inc. St. Clair, Michigan. Net Weight – 10 lbs. Darling picture of boy putting salt on bird's tail. $15-20.

Medium Salt Right & Ready RR. The letters look like people. 100 lbs. Riggles & Rademaker Salt Co. Manistee, Michigan NRA Member US, picture of eagle. We Do Our Part. $40-50.

Kiln Dried Chippewa Granulated Salt. The Ohio Salt Co. Wadsworth Ohio. 100 lbs. Net when packed. Great Indian picture. $50-65.

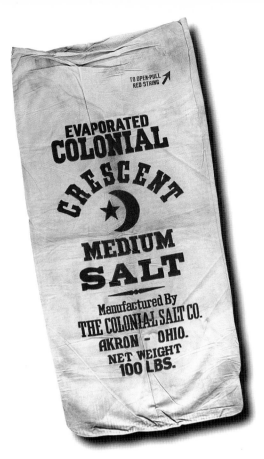

Evaporated Colonial Crescent Medium Salt. Manufactured by the Colonial Salt Co. Akron, Ohio, Net Weight 100 lbs. $40-50.

Sterling Salt, International Salt Company, Incorporated Avery Island, Louisiana. $30-45.

50 lbs. Net Weight Silver Springs Brand Coarse Salt, Worcester Salt Co., New York, N. Y. Manufacturer. $25-30.

Salt for Meat Curing. This bag has holes but is so interesting I wanted to share it. Distributed by Diamond Crystal Salt Co., Inc. St. Clair, Michigan. Made in USA. $35-40.

Rice Sacks
Depending on the type, can be a consumable or a seed.

Quick and Easy Riceland Enriched Rice. Extra Long Grain Stuttgart, Arkansas 100 lbs. $35-40.

Bon Ton Long Grain Rice 10 lbs. Milled By Estherwood Rice Sales, Elton, La. Louisiana Agriculture Products. $10-15.

Excel 100 Enriched Extra Long Grain Rice Carlson Rice Mill, West Memphis, Ark. Net Wt. 100 lb. Foreign language around edge probably means for export. $40-50.

Pacific Pride U. S. No. 1 Extra Fancy Pearl Rice, Pacific International Rice Mills, Inc. San Francisco, California 100 lbs. Net Weight. Great picture. $45-60.

Seed Sacks

Agricultural seeds of so many types. The cloth or fabric of these bags is usually very coarse and heavy

Lifetyme Brand Mixed Grass Seed Behnm & Hagemann, Inc., Peoria, Ill. Net Wt. 5 lbs. $5-7.50.

Towel Stripe Seed Bag S. B. Simons & Sons, Elkhorn, Wis. Member of the Wisconsin Experiment Association. Picture of Wisconsin map, corn cobs, and corn shocks. Hybred Seed Corn, Soybeans, and Vicland Oats. $30-35.

Fulton Seamless An extra heavy side stripe bag. $40-45.

Kentucky 31 Fescue The Wonder Grass Produced by G. W. Jones & Sons Farms Huntsville, Alabama. 50 lbs. Net $25-30.

Medium Red Clover 60 lbs. Net Weight. $20-25.

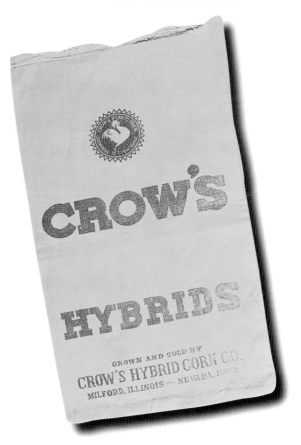

Crow's Hybrids Grown and Sold By Crow's Hybrid Corn Co., Milford, Illinois – Nevada, Iowa, Something to Crow About! $20-25.

Harpel's Hybrids Your Shield of Assurance Corn King Years 1947-1956, Walter J. Harpel Seed Co, Crawfordsville, Indiana, Modern Plant Shannondale, Indiana. This is a local bag to my community and therefore would sell at a higher price in my area. $50-75.

D & H Seeds 1 Bu. Farm Seeds DeWine & Hamma Yellow Springs, O. Semmes Bags. $20-25.

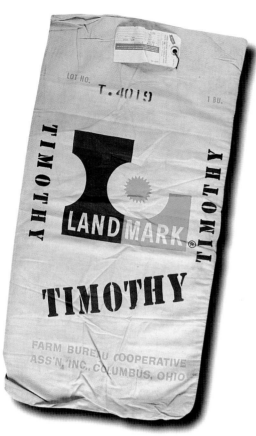

Landmark Timothy Farm Bureau Cooperative Ass'n. Inc. Columbus, Ohio. $20-25.

Nebr. Alfalfa, Noble Brand Trade Mark Seeds, One Bushel, Packed by Noble Bros., Seed Co. Gibson City, Ill. $20-25.

Adapted Seed of Selected Origin. Farm Bureau Co-op Indiana Farm Bureau Cooperative Assn., Inc. One bushel when packed. $15-20.

Farmelco Seeds, One Bushel 60 lbs. Net Weight, Farmers Elevators of Illinois. Packed for Illinois Grain Corporation, Farmelco Co. Supply Division Chicago, Illinois. $15-20.

Funk Bros. Seed Co. Bloomington, Ill. Always
No. 1 Grade One Bushel. $20-25.

Certified Ranger Alfalfa Seed for Profitable
Farming 60 lbs. Net Wt. $20-25.

Landmark Certified Vernal Alfalfa, Farm Bureau Cooperative
Ass'n., Inc. Columbus, Ohio. $20-25.

Certified Ranger Alfalfa Seed Net Weight 60 lbs. Fulton Bag. $20-25.

Ladino Clover Certified Seed 50 lbs. Net Wt. $15-25.

Dickinsons Pine Tree Farm Seed. This is a very tough textured bag with stripes. $20-25.

Farm Feeds And Supplements

I am so glad I included these in my collection as I am simply falling in love with their history. I guess it is my farm background and child hood.

Diamond Sunshine Egg Mash Manufactured by D. P. Wigley Co, Racine, Wis. Chase Bag Co. 100 lbs. Net. Picture of hen and two chicks in large diamond. $45-55.

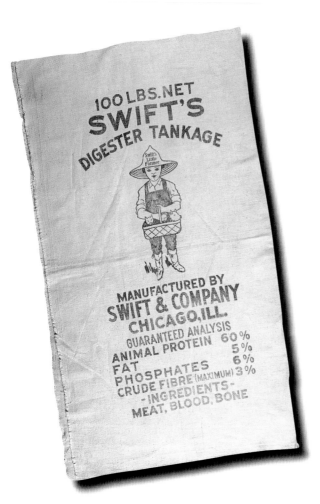

Swift's Digester Tankage Manufactured by Swift & Company, Chicago, Ill. Picture of a boy in stripe overalls with boots and hat and carrying basket with three baby pigs. $50-60.

Red W Brand Pig pictured Digester Tankage Meat Meal Guaranteed Analysis, Wilson & Co. Chicago, USA. $45-50.

Blue Label Feed Calf Starter. Cow with two calves and barn. Country Road Calf Starter. Moran Feed & Seed Co. Est. 1880. $45-50.

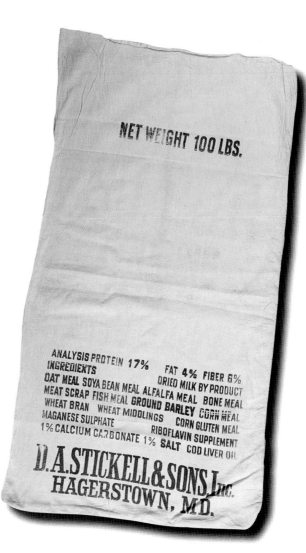

D. A. Stickell & Sons, Inc. Hagerstown, Md. Ingredients of feed listed. $40-45.

Felco Quality Feeds, Balanced Rations for every purpose. Back of bag reads: "The name Felco on a bag of feed guarantees you the best value that your own cooperative can put into it." $40-45.

Beacon Battery, The Beacon Milling Company, Inc. Beacon Feeds Light the Way to Better Feeding, Cayuga, New York. Poultry Growing Ration. Chase Bag Company with instructions for removing print. 100 lbs. $45-60.

Ceres Lay Mash, large Ceres Logo and Chicken Head. Ceres Supply Co, Inc. Massillon, Ohio, 100 lbs. $40-45.

Wayne Feeds 26% Supplement Egg & Breeder Manufactured By Allied Mills, Inc. 26% Supp. Picture of man on horse. 100 lbs. $45-50.

New Idea 32% Supplement Mash Less Cost Greater Profit. Manufactured By Crabbes Reynolds Taylor Co. Analysis Guaranteed. $40-45.

Feed Purina Right Purina Hog Chow Supplement. Checkered red center, blue squares. 100 lbs. $40-45.

Purina Broiler Chow. Contains Puri-Flave Vitamin – G Flavin and Pur-a-tene (Carotene). Feed Purina Right. 100 lbs. $40-45.

Arcady Feeds for all Live Stock and Poultry. Made by Arcady Farms Milling Co., Chicago, Ill. Five Point Protection, Copyright 1941 by Arcady Farms Mfg. Co. "There is an Arcady feed for every farm need." 100 lbs. $40-45.

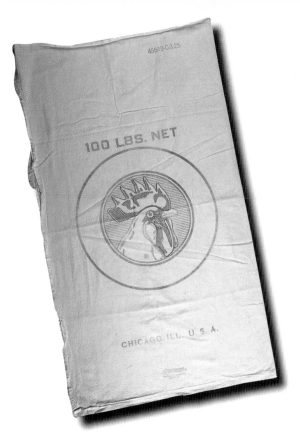

Chicken head in circles. 100 lbs. Chicago, Ill. USA Central Bag Co. $40-45.

Horn of plenty picture with horses, pigs, cow, chicken, geese and others. Horn's Diamond Feeds D. E. Horn & Co., York, Pa. $40-45.

Someone used the front of this sack and quilted it. Eshelman Red Rose Poultry Feed. Picture of hen and two chicks. John W. Eshelman & Sons Established 1842. 100 lbs. $45-50.

Red Rose Poultry Feed. Large pictures of chickens and red roses. John W. Eshelman & Sons, Lancaster, Pa./York, Pa./Circleville, O./Tampa, Fla./Sanford, N. C./Chamblee, Ga. $50-55.

Beacon Broiler Feed with pictures of beacons. Manufactured by The Beacon Milling Co., Inc. Cayuga, New York. 100 pounds. $55-75.

Ful-O-Pep Poultry Litter. Two red chickens shown. Made by the Quaker Oats Company. 50 lbs. Net Weight. The design indicates possible fighting chickens, which would not be a desirable sport. Chicken pictures are very desirable, however, leading to a higher value for this bag. $75-100.

Anderson Quality Scratch Grains Manufactured by Anderson Grain & Feed Co. York, Pa. Decorated with red letter "A's" all around the edge of the bag. $40-45.

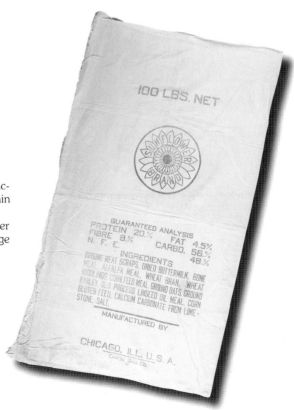

Sunflower Brand, flower picture. 100 lbs. Net – Ingredients listed. Chicago, Ill. USA. Chase Bag Co. $40-45.

Olympic Egg Mash 100 lbs. Sperry Flour Company Trade Name of General Mills, Inc. Offices San Francisco, Cal. Pellets. $40-45.

Red Comb Poultry Feeds, Red Comb Crumbles. Picture of chicken head. Hales & Hunter Co, Chicago, Ill. U.S.A. HLCE Ribbon Bags Central Bag & Burlap Company, Chicago, Ill. $45-50.

Picture of three calves, Cooperative Mills, Inc. Quality Feed & Grain Milton, S. C. Enriched with Vitamins and Iron 100 lbs. Net. Up in the corner in red print over a chicken picture the words "Freight Shipping Bag Guaranteed by Hyland Bag. Meeting requirements of Con-Freight Classification for grain products." $40-45.

Race Horse with jockey picture. Fancy Rex oats,
Premium Quality Race Horse Oats. ADM Archer
Daniels Midland Company, Minneapolis, Minnesota. 98 lb. bag. $45-50.

Purity Windmill sign picture. The Right Sign to Follow.
Mash Supplement 100 lbs. Net. The Urban Mills Co.
Urbana, Ohio, Werthan. $40-50.

Stonemo Crushed Granite Insoluble, Stone
Mountain Grit Co., Lithonia, Ga Hen Case
Bag Co. $30-35.

Made-Rite Feeds, Linn County Grain Co, Cedar Rapids, Iowa. Animals pictured on star and circle. $35-40.

Haynes, Statepilot 10 lbs. Net New Mashettes. Manufactured by Haynes Milling Co., Inc. Portland, Indiana. $40-45.

100 lbs. Net New Ross All Mash Chick Starter and Grower. New Ross Grain & Lumber Company, New Ross, Ind. This is a local bag to my community and would therefore hold more value in my area. $75-100.

Pig picture 100 lbs. Net. Red W Brand Digester Tankage Meat Meal Wilson & Co., Chicago USA. $40-45.

Polar Bear Gray Shorts. Made from pure wheat by the New Era Milling Co., Arkansas City, Kansas. 100 lbs. $50-75.

Topper 16% Dairy Feed Guaranteed by Allied Mills, Inc., Chicago, Ill. 100 lb. $40-45.

Wonderfat Feeds Arcady Farms Milling Co., Chicago, Ill. N. Kansas City, Mo. Red chickens all around edge. Note the re-stitch on the edge in blue thread. Many bags were re-stitched and holes repaired in order to continue using the bags over and over. $45-60.

Bossie Stock Feed Manufactured By Lexington Roller Mills, Inc. Lexington, Kentucky. Stock Feed 100 lbs. Net Wt. Cow and pig picture. $40-45.

Sugared Schumacher Feed Oct. 17, 1911. Made by The Quaker Oats Company Address Chicago, USA 100 lbs. Net Weight. $40-45.

Cow's head picture. Sugarine 17% Dairy Feed. Guaranteed by Allied Mills, Inc. Chicago, Ill. $45-50.

Pillsbury's Best Feeds Laying Mash. Paper label on green, orange, blue, and red striped bag. Pillsbury Mills, Inc. Feed and Soy Division, Clinton, Ia, Louisville, Ky., Los Angeles, Calif. $65-85.

Good-Rich Poultry. Animal pictures. Concentrate Fortified
with A-Factory Goodrich Feed Mills, Winchester, Indiana,
Copyright 1948. $35-40.

Kingan's Digester Tankage Hog Food Kingan & Co.
Pork & Beef Packers, Indianapolis, Ind. $25-30.

Superior Balanced Egg Pellets,
Oklahoma City, USA. $45-50.

Blue and red ink printing on yellow bag with brown and yellow geometric designs. Wayne Egg Mash Guaranteed by Allied Mills, Inc., Chicago, Ill. $45-65.

Ranch Way Chick Scratch, Denver, Colorado. The Colorado Milling & Elevator Milling & Elevator Co. Ranchway Division. R. W. Fulton Band – Label. "To remove band label simply soak in water." 100 lb. bag. Purple and lavender flowers with blue squiggles in between. $45-65.

Ranch Way 100 lbs. For Poultry and Livestock Feeds Manufactured by the Colorado Milling and Elevator Co. Fulton Brand – Label bag. "To remove band label simply soak in water." Pink and blue flowers print. $45-65.

Lay or Bust Feed. Shows picture of a hen exploding.
Dry Mash Manufactured by the Park & Pollard Co. of
Ill. Chicago, Ill. Bemis Brothers Bag Co. $45-55.

Green/white diagonal plaid printed feed sack. Protein Blend-
ers, Inc. Iowa City, Iowa. Feed B P Blended Protein. Wertman
Banded Bag. "To remove label soak in water." 100 lbs. Pig
Concentrate. $55-75.

Jim Dandy printed striped bag, paper band – label. Laying Mash a Breeder Mash. Bag is Bemis Band Label bag.
"Soak in water to remove print and label." Print on bag
directly and also paper label with print. $35-45.

Paper banded geometric pink and white bag. Bang-Up Dairy
Feed. Manufactured by the Jim Dandy Co., Birmingham,
Ala. 39 x 55 Bang Up 24% Dairy Feed c 1959. $60-75.

Chicken Feed Sacks With Prints Like Sewing Fabric

My favorite of all time. Thank you for saving this wonderful world and history of chicken feed sacks. The pictures follow and I hope you enjoy. There is one category we failed to include, as we simply overlooked them at the photo shoot. That is the paisleys. So please forgive me for missing those cute little comet shaped prints. We have pictured novelty, geometric, pillowcase, floral, border, fruit, and solid color categories.

Novelty Print Chicken Feed Sacks

Novelty prints generally consist of the following pictures: children's prints, kitchen prints, people, pictures of houses, or simply pictures of any objects or scenes. Classic stories or nursery rhymes, themes, movies, current events such as wars, historical events, western prints and Mexican or Southwestern prints are also novelty prints. One might consider bows, ribbons, birds, butterflies, and many other objects that are also included in basically floral sacks as novelty prints. Some of the most expensive would be the Disney characters.

What I would not do to find the Buck Rogers, Gone with the Wind, Bo Peep, Humpty Dumpty, Cow Boy Prints, and Mickey Mouse! I have seen some in the past that were doll patterns to be cut out and stuffed to make dolls. The price thought to be too stiff then probably is not too high today. No, I do not have a Mickey Mouse but I figure it to be in the $350 and up range.

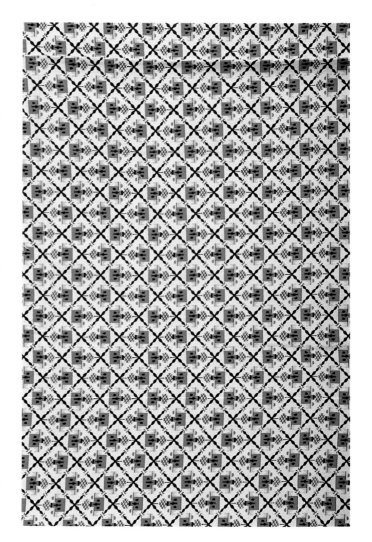

Green and yellow houses in black border diamonds. $40-45.

Southwest print with large tree like cactus, children groups, ladies with fruit bowls on heads, men playing guitars. $35-45.

Long ribbons with bows at slants. $30-35.

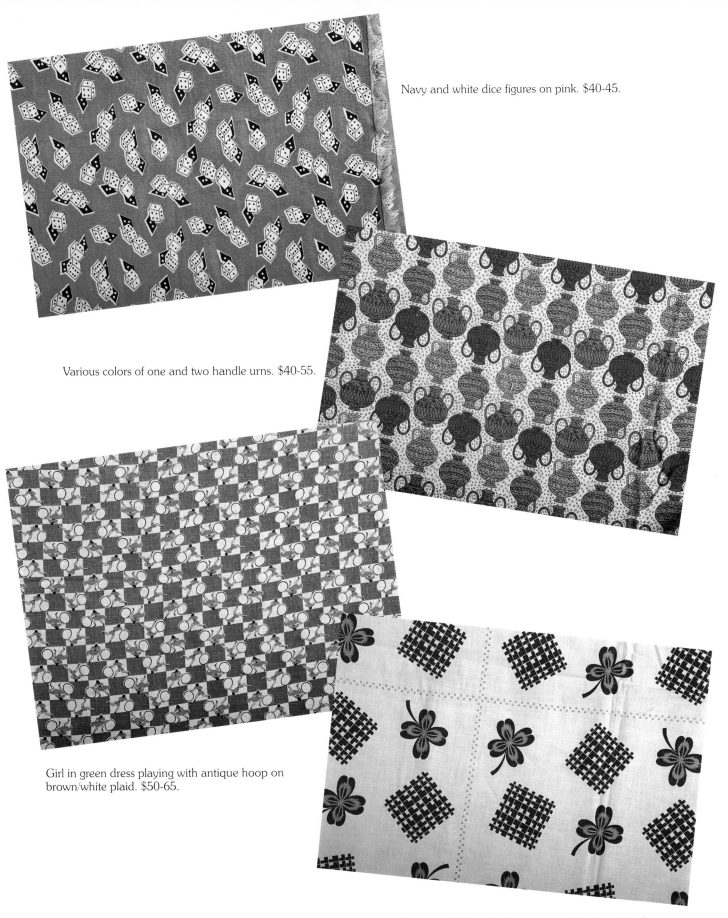

Navy and white dice figures on pink. $40-45.

Various colors of one and two handle urns. $40-55.

Girl in green dress playing with antique hoop on brown/white plaid. $50-65.

Blue four leaf clovers and blue plaid squares, smaller size print of other bags with this same design only larger in size. I see many bags that have a larger and small size print but same design and in different colors. $30-35.

59

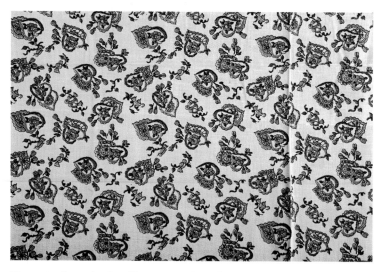

Turquoise/lavender floral hearts on white. $30-35.

Red, navy, and white border print with trees, flowers, boats, houses, buds, shepherds and deer. $55-60.

Red four leaf clovers and red plaid squares with red clover border. $35-40.

Turquoise/white small plaid with yellow windmills. $40-55.

Women with jugs on head, women kneeling, cactus and palms, women cooking leavened bread. $35-55.

Blue roses with yellow bell wreaths and yellow hats with blue bands and pink bows, pink ribbon and bows. $30-40.

Flower packages or cubes in green, yellow, blue, and red. $30-35.

Jugs, plates, strawberry pitchers, dishes on shelves covered with green apple shelf paper. $40-55.

Green teapots, yellow cups, green saucers, yellow creamer and green sugar. $35-45.

Blue, yellow, orange bows among ball flowers. $30-35.

Green and white palms with navy and green dots. $30-45.

Blue and green mushrooms, blue flowers, and red birds. $35-50.

Brown palm trees among orange, turquoise, and green scrolls. $40-45.

Navy palm trees among yellow, gray, and orange scrolls. $40-45.

Red, pink and yellow large letters. $35-40.

Large green four leaf clovers and plaid squares on border print. $40-45.

Brown thatched among a jungle of palms. $45-55.

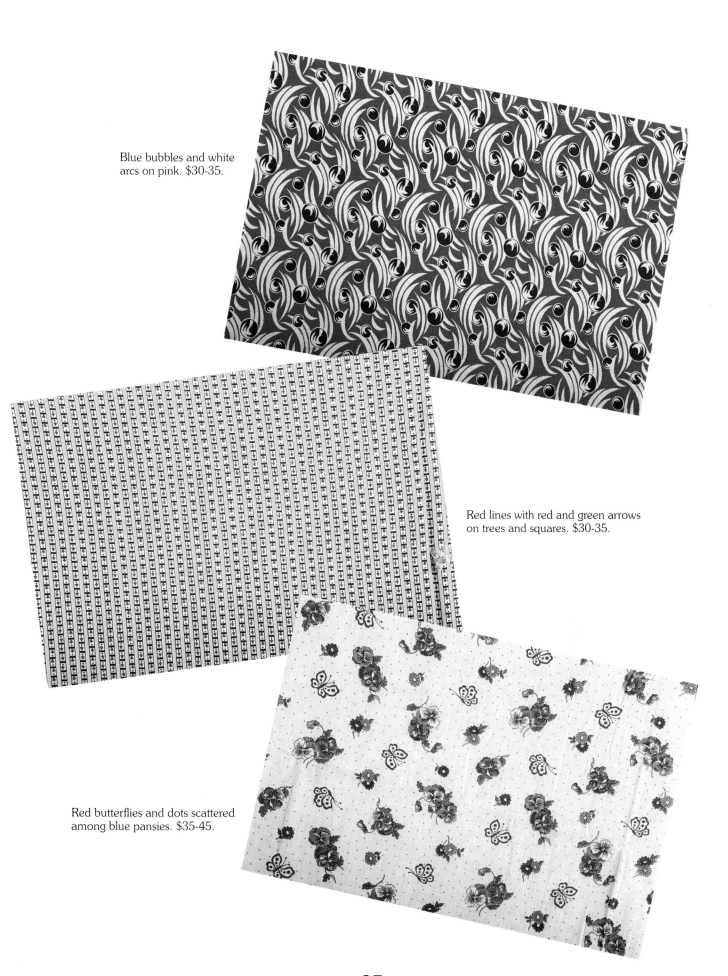

Blue bubbles and white arcs on pink. $30-35.

Red lines with red and green arrows on trees and squares. $30-35.

Red butterflies and dots scattered among blue pansies. $35-45.

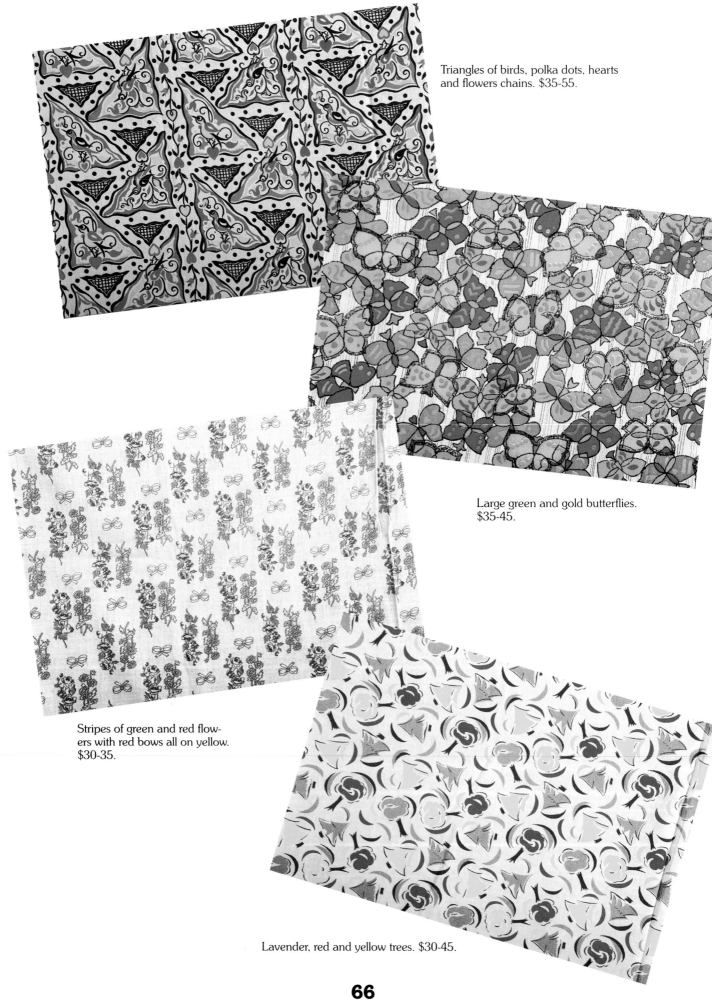

Triangles of birds, polka dots, hearts and flowers chains. $35-55.

Large green and gold butterflies. $35-45.

Stripes of green and red flowers with red bows all on yellow. $30-35.

Lavender, red and yellow trees. $30-45.

Blue with pink baby animals, chicks, deer, teddy bears, elephants, and bears flying kites. $55-75.

Red and lavender feathers scattered. $30-35.

Red, yellow and black stick figures with Christmas trees and logs. $30-35.

Purple heart flower pots holding tulips on white. $30-35.

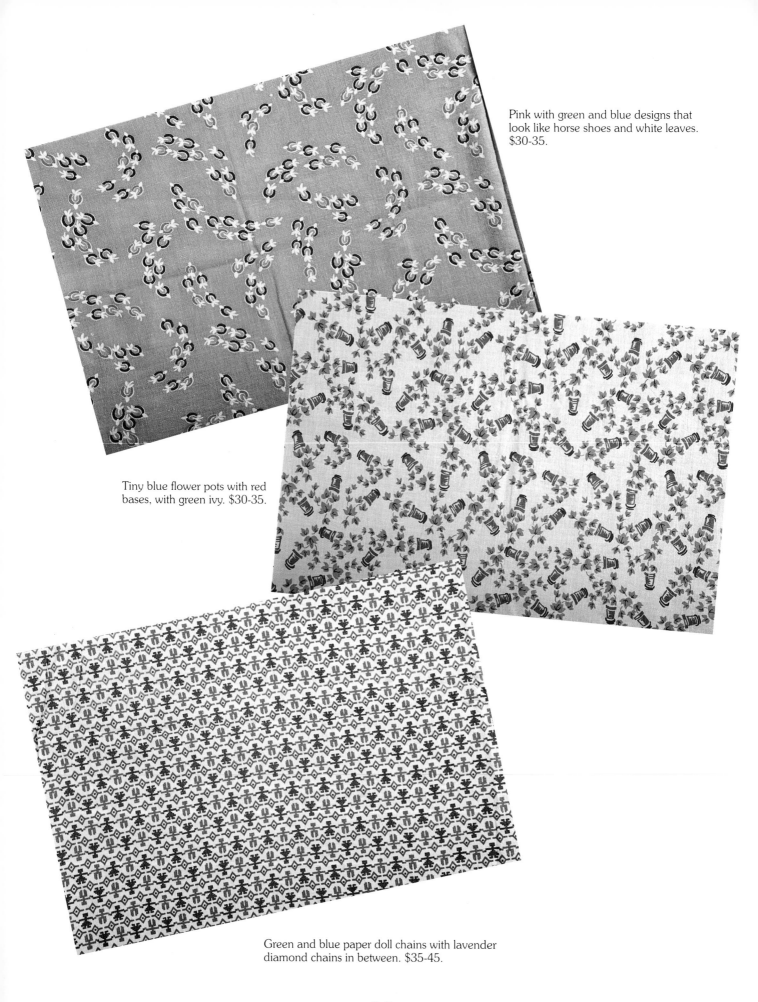

Pink with green and blue designs that look like horse shoes and white leaves. $30-35.

Tiny blue flower pots with red bases, with green ivy. $30-35.

Green and blue paper doll chains with lavender diamond chains in between. $35-45.

Nautical print with blue ship wheels, life preservers connected by net or rope. $35-40.

Horse heads or chess knights, brown and orange on white. $35-40.

Navy blue and white three leaf clovers in squares. $30-35.

Bright blue, red and green chickens, with light blue scattered cactus, mandolins, sombreros, and jugs. $75-85.

Green and yellow umbrellas with white flowers. $35-45.

Small 1" square pictures of southwest siesta, hanging peppers, cactus, ladies with braided hair with colorful pottery, ladies cooking, men playing mandolins, cactus and pots. $45-65.

Large southwest dressed characters in lavender , grey, yellow and green with red palm trees, cactus, mandolins, sombreros, and pottery. $40-55.

Yellow flower carts with red and yellow tulips and tiny lavender flowers. Streamers and flowers are between the carts. $35-55.

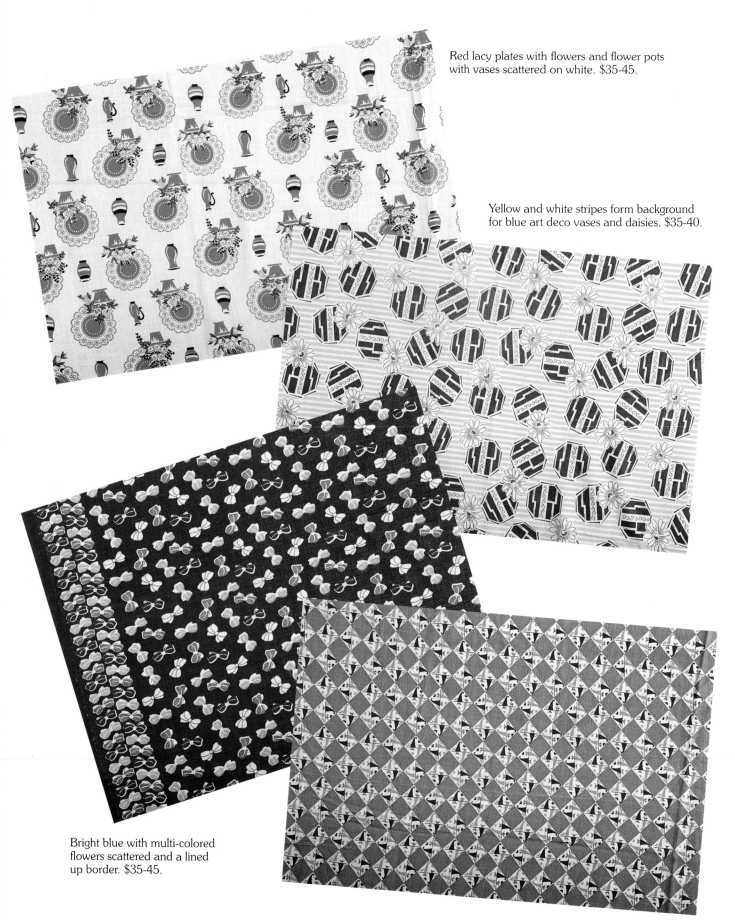

Red lacy plates with flowers and flower pots with vases scattered on white. $35-45.

Yellow and white stripes form background for blue art deco vases and daisies. $35-40.

Bright blue with multi-colored flowers scattered and a lined up border. $35-45.

Turquoise and white diamonds with brown and green wild-looking birds on boats. $30-40.

Tiny Indian print in lavender, turquoise and black on white. Horses and tee-pees, horses, and buffalos. $60-75.

Tiny Indian print in lavender, turquoise and black on white. Horses and teepees, horses, and buffalos. $60-75.

Red plaid with scattered blue butterflies. $30-35.

Green picture frames on red. Inside the frames are bouquets with bells. $35-45.

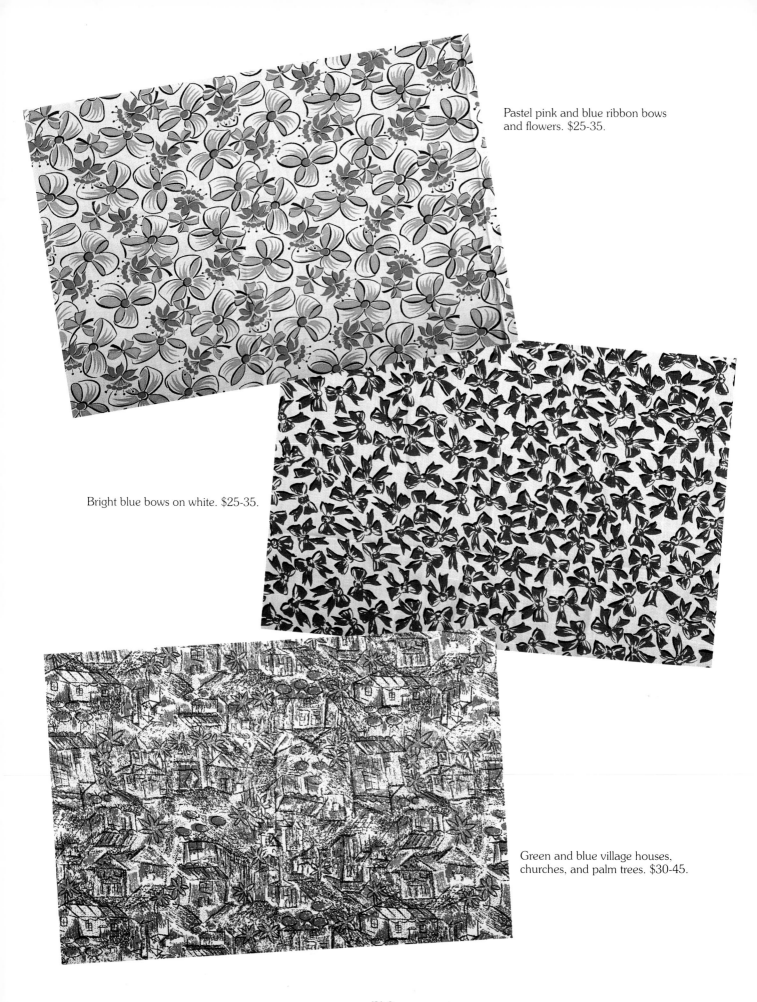

Pastel pink and blue ribbon bows and flowers. $25-35.

Bright blue bows on white. $25-35.

Green and blue village houses, churches, and palm trees. $30-45.

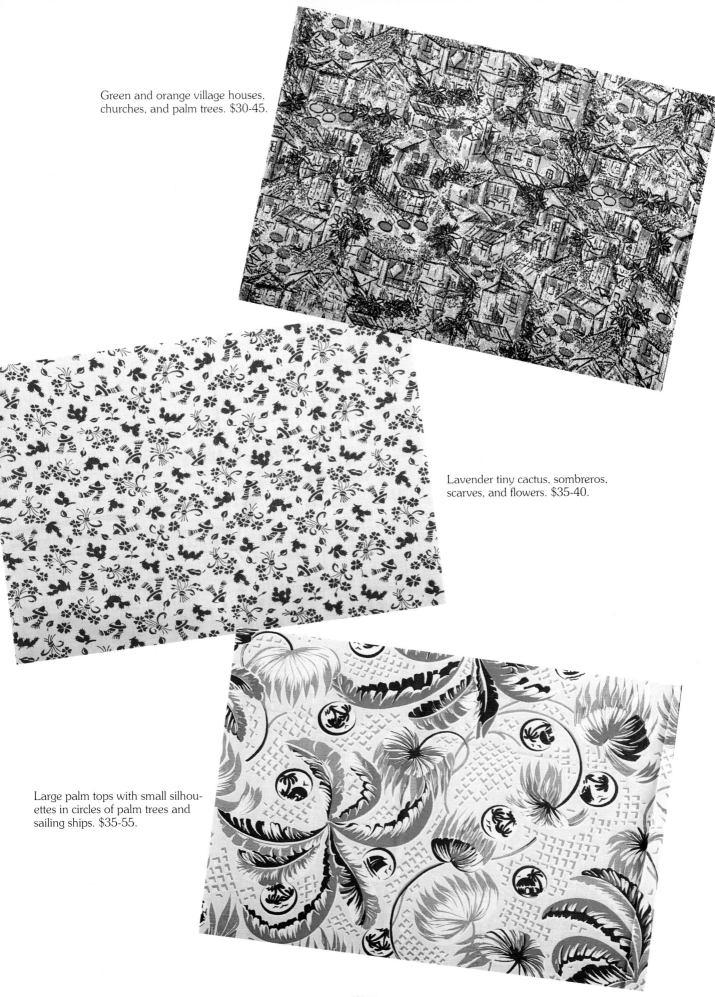

Green and orange village houses, churches, and palm trees. $30-45.

Lavender tiny cactus, sombreros, scarves, and flowers. $35-40.

Large palm tops with small silhouettes in circles of palm trees and sailing ships. $35-55.

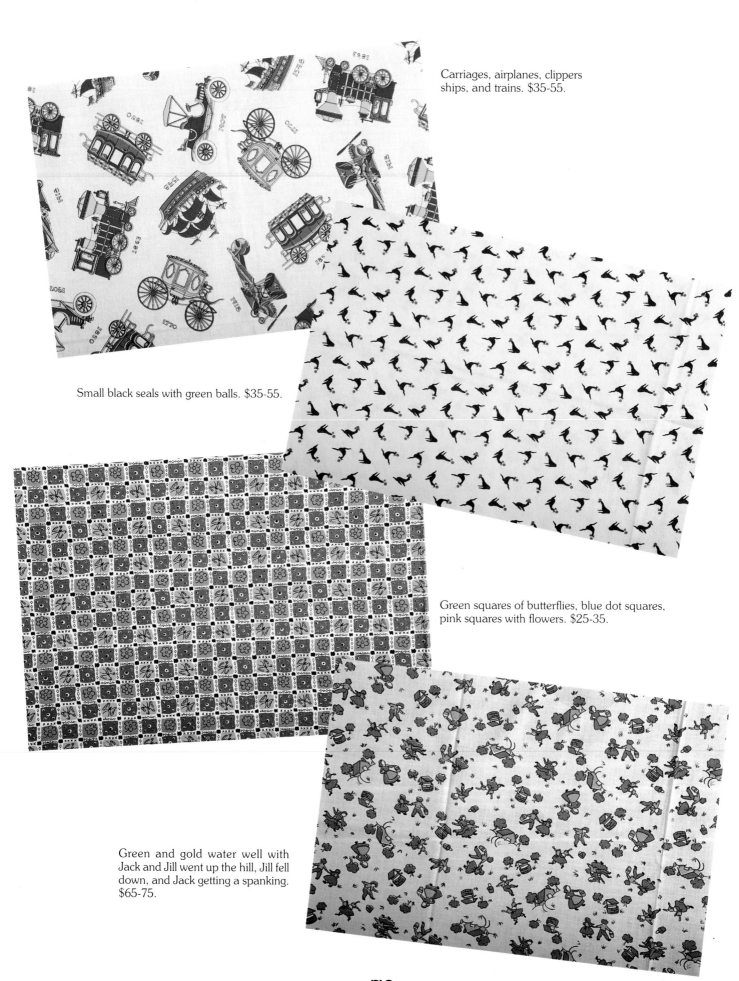

Carriages, airplanes, clippers ships, and trains. $35-55.

Small black seals with green balls. $35-55.

Green squares of butterflies, blue dot squares, pink squares with flowers. $25-35.

Green and gold water well with Jack and Jill went up the hill, Jill fell down, and Jack getting a spanking. $65-75.

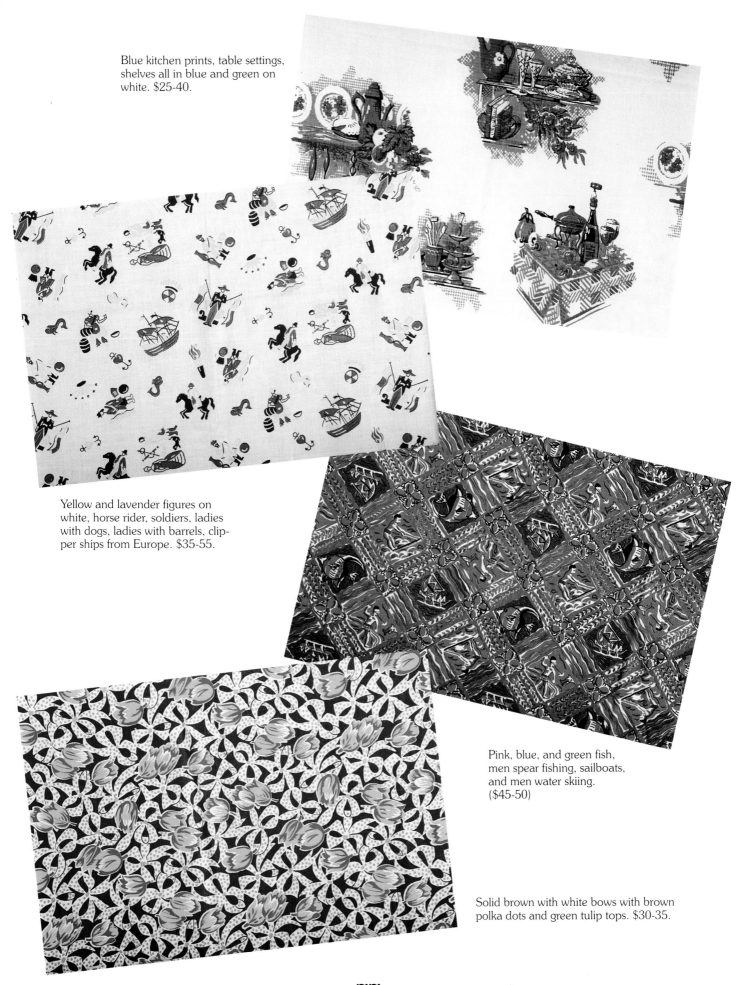

Blue kitchen prints, table settings, shelves all in blue and green on white. $25-40.

Yellow and lavender figures on white, horse rider, soldiers, ladies with dogs, ladies with barrels, clipper ships from Europe. $35-55.

Pink, blue, and green fish, men spear fishing, sailboats, and men water skiing. ($45-50)

Solid brown with white bows with brown polka dots and green tulip tops. $30-35.

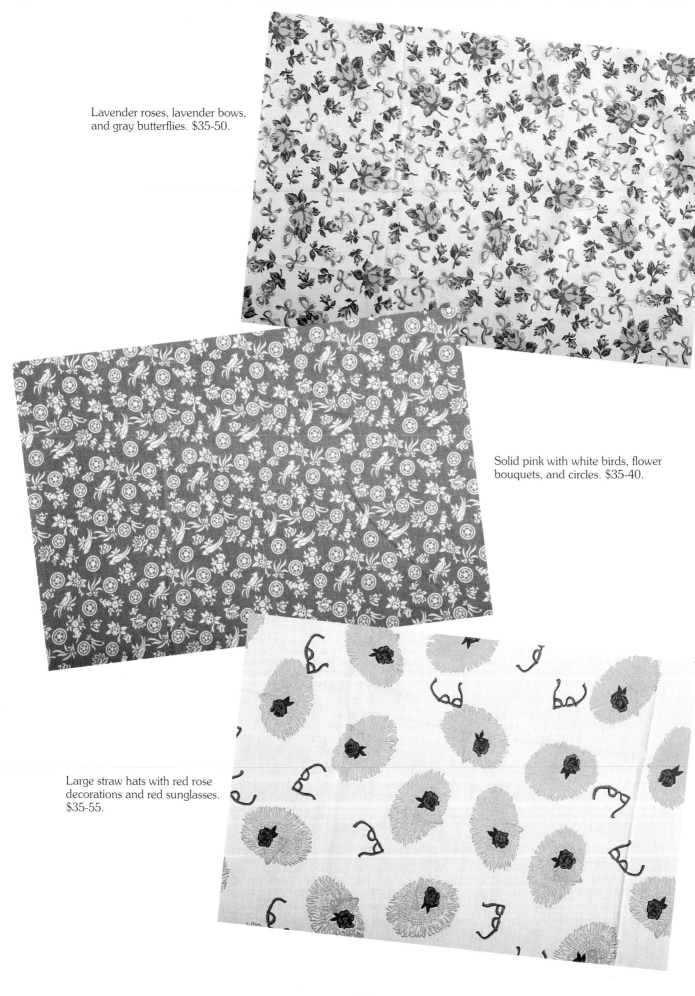

Lavender roses, lavender bows, and gray butterflies. $35-50.

Solid pink with white birds, flower bouquets, and circles. $35-40.

Large straw hats with red rose decorations and red sunglasses. $35-55.

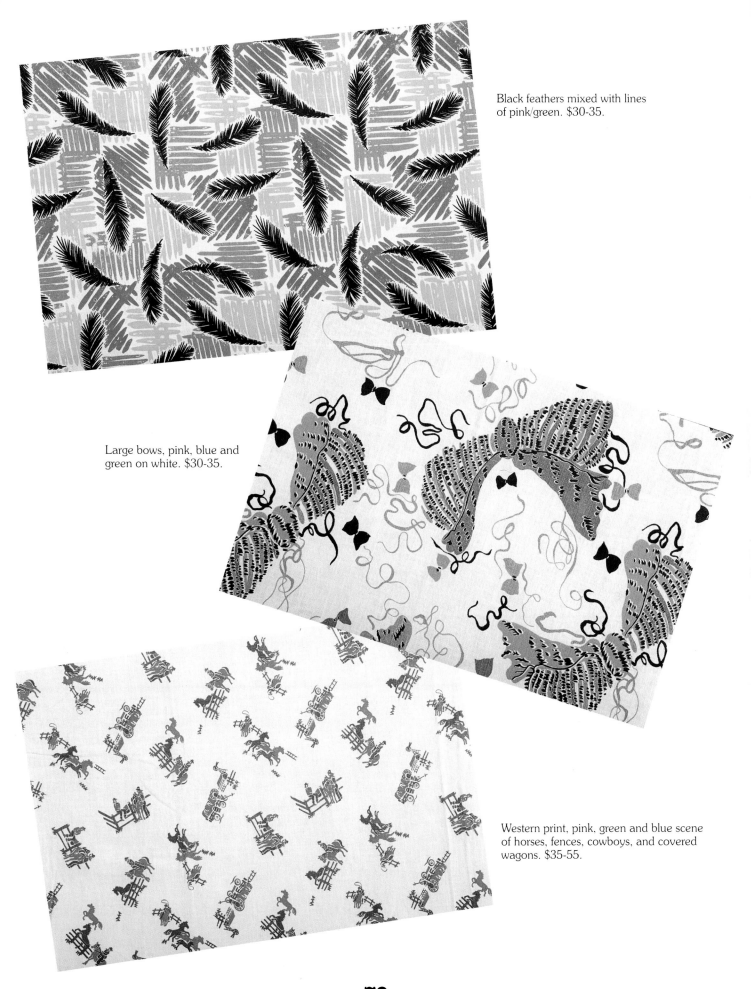

Black feathers mixed with lines of pink/green. $30-35.

Large bows, pink, blue and green on white. $30-35.

Western print, pink, green and blue scene of horses, fences, cowboys, and covered wagons. $35-55.

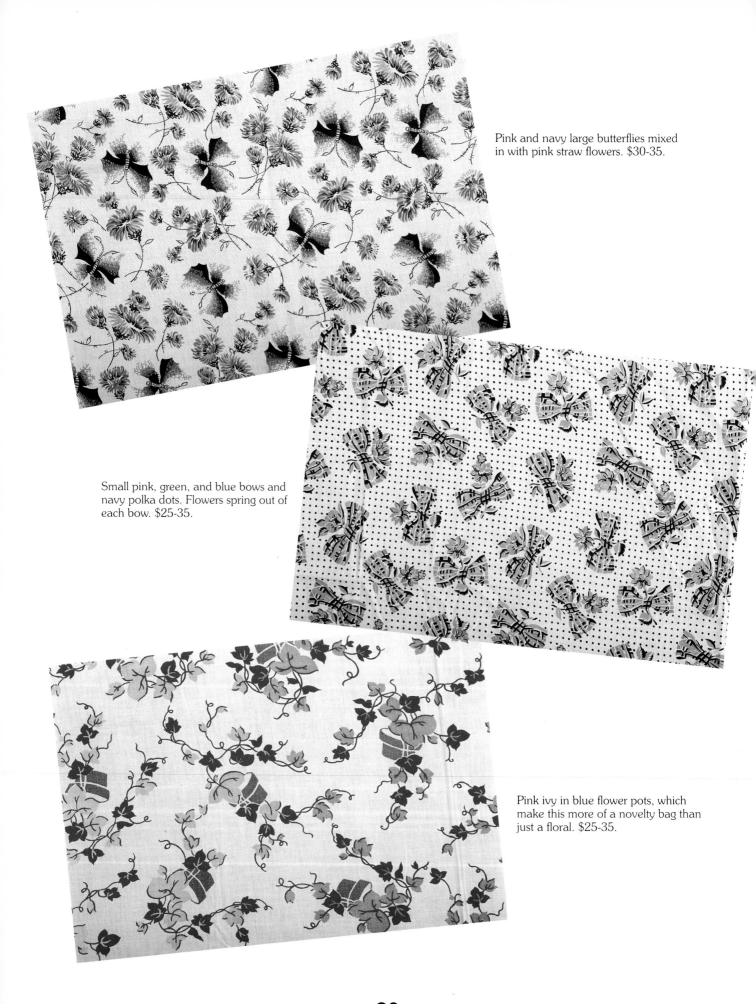

Pink and navy large butterflies mixed in with pink straw flowers. $30-35.

Small pink, green, and blue bows and navy polka dots. Flowers spring out of each bow. $25-35.

Pink ivy in blue flower pots, which make this more of a novelty bag than just a floral. $25-35.

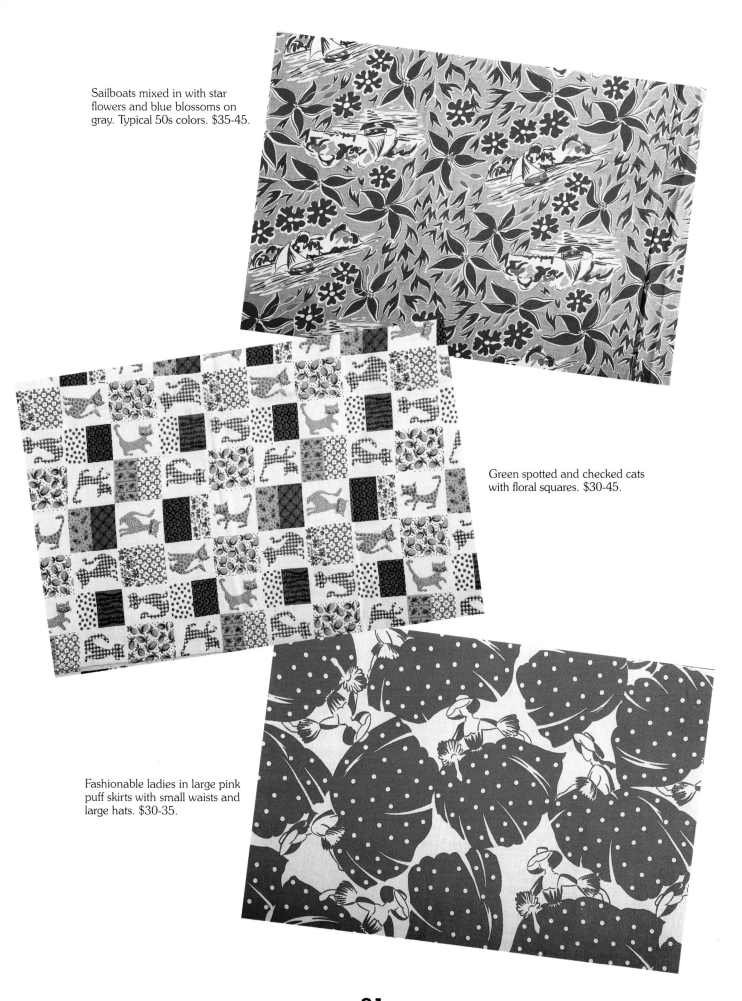

Sailboats mixed in with star flowers and blue blossoms on gray. Typical 50s colors. $35-45.

Green spotted and checked cats with floral squares. $30-45.

Fashionable ladies in large pink puff skirts with small waists and large hats. $30-35.

Red plaid with navy silhouettes of children flying kites, jumping rope, walking to school with books with a dog, and walking in rain with umbrellas. $45-60.

Rust and aqua flying geese on yellow. $30-45.

Red and green pineapples on plants and palm trees. $40-45.

Black and brown leopard spots on white. $35-50.

Black and white zebra stripes. $35-50.

Fishermen with large nets, sailboats, peasant ladies with boys and mixed with red poppies. $50-75.

Peasants with jugs, turret style buildings, cactus, palms, and apple trees or red fruit. $45-60.

Farm sleigh scene with barns and snow. $35-40.

Dutch men and women with blue swans with red dots between. $45-60.

Red flowers with mandolins on solid light blue. $45-50.

Native girl surfing with beach scenes of green and yellow palm trees. $45-55.

White hearts with tiny pink and blue flowers on navy. $35-40

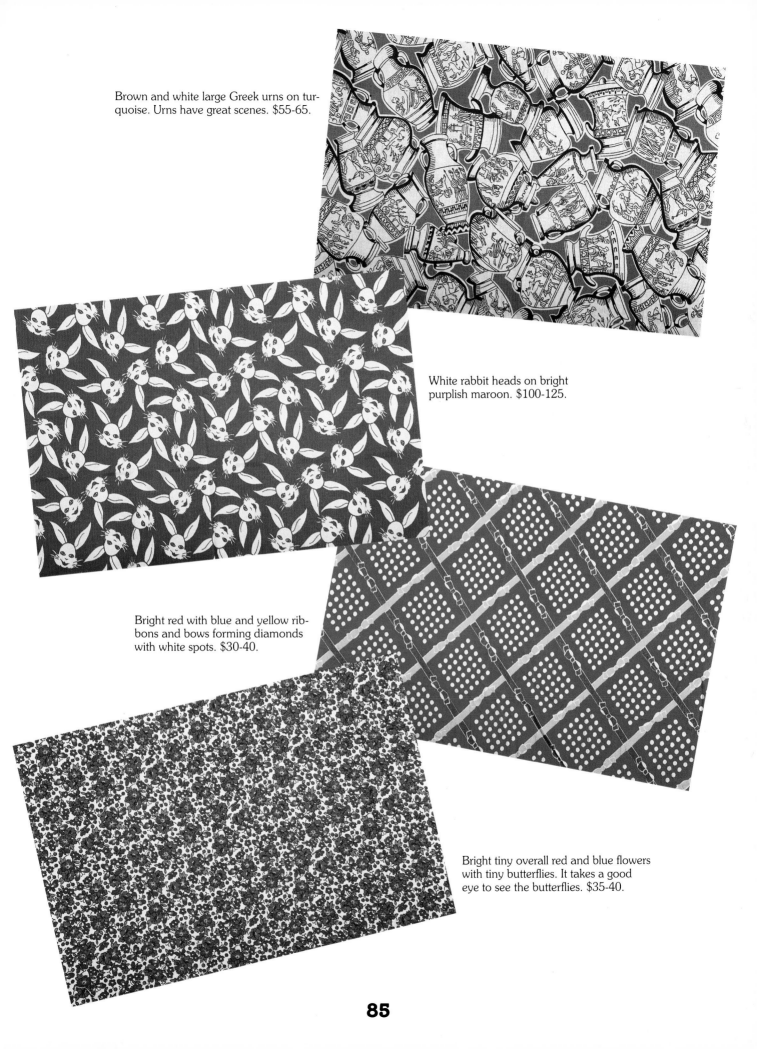

Brown and white large Greek urns on turquoise. Urns have great scenes. $55-65.

White rabbit heads on bright purplish maroon. $100-125.

Bright red with blue and yellow ribbons and bows forming diamonds with white spots. $30-40.

Bright tiny overall red and blue flowers with tiny butterflies. It takes a good eye to see the butterflies. $35-40.

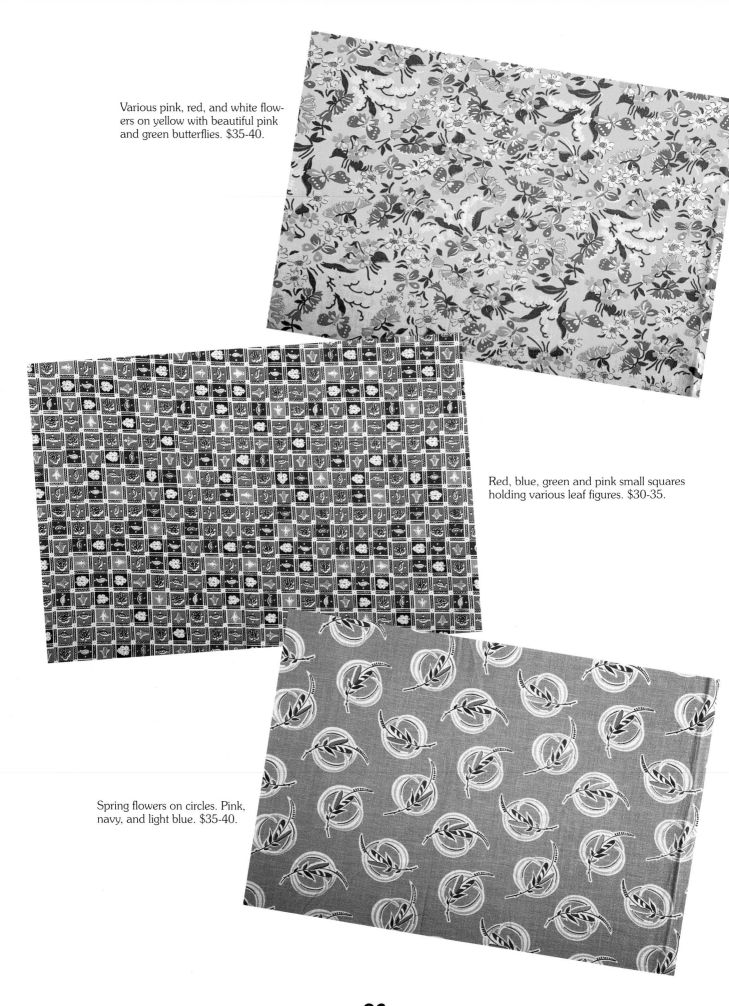

Various pink, red, and white flowers on yellow with beautiful pink and green butterflies. $35-40.

Red, blue, green and pink small squares holding various leaf figures. $30-35.

Spring flowers on circles. Pink, navy, and light blue. $35-40.

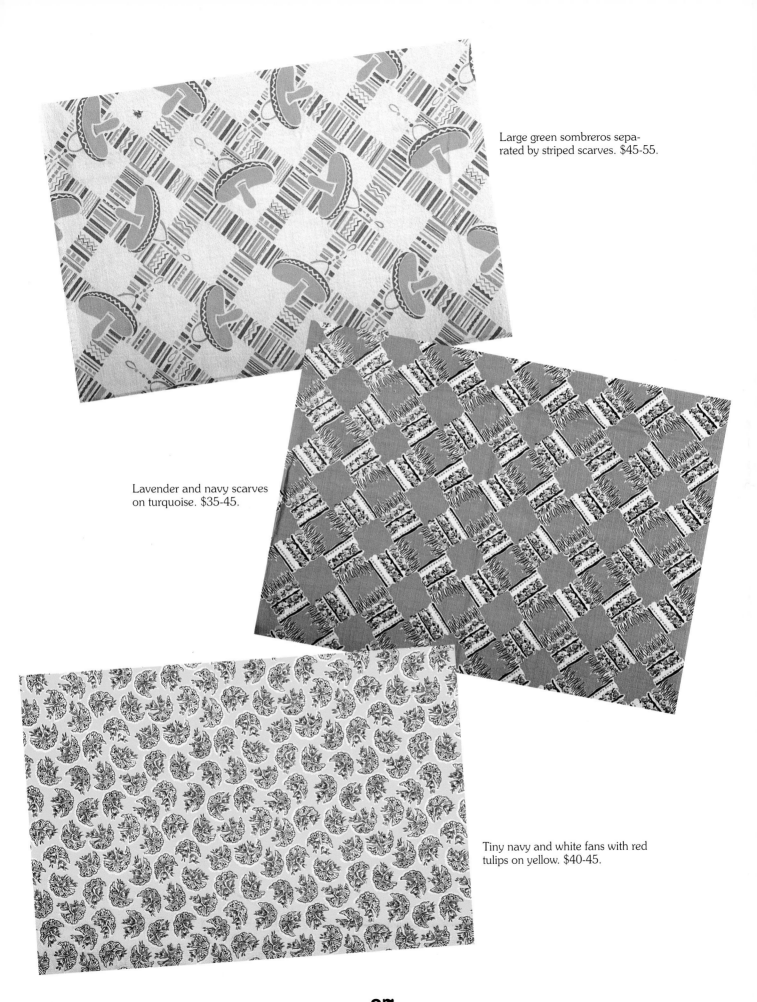

Large green sombreros separated by striped scarves. $45-55.

Lavender and navy scarves
on turquoise. $35-45.

Tiny navy and white fans with red
tulips on yellow. $40-45.

Large fancy blue and white fans on red with white snowflake-like geometric figures. $40-45.

Navy antique cars with white scrolls on light blue. $45-50.

Tiny teepees red, yellow and green on white. $45-60.

Merry-go-rounds of blue, red, and green with same colored ribbons and bows between all on yellow. $45-55.

Pink and lavender rodeo and cowboy scenes. Lots of fences. $45-60.

Navy framed silhouette heads on blue. $45-55.

Brown framed silhouette heads on gold. $45-55.

Southwest children and donkeys in brown and blue on yellow. $45-55.

Tropical girls water skiing, picking flowers, huts, pink, green and brown palms on white. $35-55.

Pink/white and yellow/white feathers on turquoise. $35-40.

Pink and blue hearts on navy with white ribbons. $35-40.

Sailboat scenes on green with brown and orange flowers. $35-45.

Navy colored men milking cows, donkey carts, hay wagons, corn shocks, farming with horses, red flower bouquets, all on solid light blue. $45-65.

Pine needles. This is a very simple pattern. $35-40.

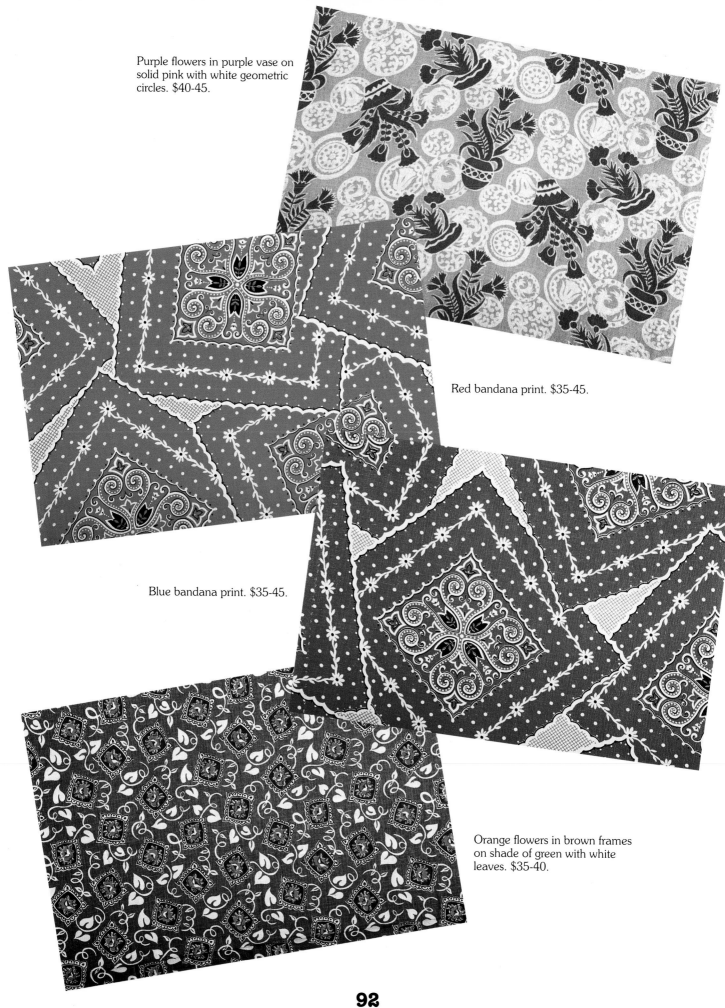

Purple flowers in purple vase on solid pink with white geometric circles. $40-45.

Red bandana print. $35-45.

Blue bandana print. $35-45.

Orange flowers in brown frames on shade of green with white leaves. $35-40.

92

Red barrels of flowers and red handled bowls. $35-40.

Peasant Dutch men and women on pink hearts. $45-55.

Gray and red rows of villages on white including schools, people, trees, bicycles, scooters, and antique cars. $45-55.

Print cross stitch large framed pictures of farm houses, square wagons, barns and water wells. $40-45.

Bright pink and blue flowers with dotted squares and bows. $35-40

Yellow baskets with pink flowers, pink tulips with green leaves. $40-45.

Shelves, scalloped shelf paper with kitchen type dishes. I have seen this print in many colors. $40-50.

Country kitchen print, wood baskets, fruit, candy jars, chicken plate, recipe books, vases, all in pink and green on white. $40-50.

Navy, pink and blue country ladies with long dresses, men with flower carts, hearts, flowers, and houses. Looks like a Pennsylvania print. $45-55.

Six sided frames holding baskets of fruit, flowers, and houses. All in red and gray. $40-45.

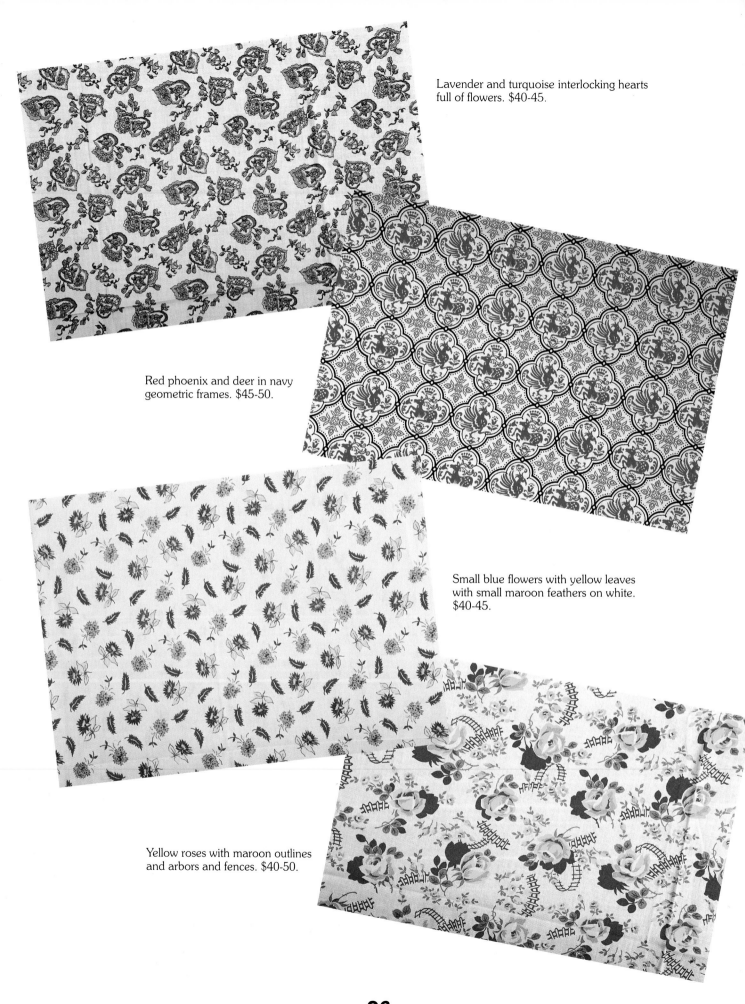

Lavender and turquoise interlocking hearts full of flowers. $40-45.

Red phoenix and deer in navy geometric frames. $45-50.

Small blue flowers with yellow leaves with small maroon feathers on white. $40-45.

Yellow roses with maroon outlines and arbors and fences. $40-50.

Navy houses separated by pink squiggles. $40-50.

Southwest design separated by blue wavy diamond frames. Red, green, blue and yellow ladies with fruit on heads, siesta men, jugs and vases, and large fruit. $45-60.

Red and pink roses framed by white lace on solid blue. $40-45.

Fancy fans framed by X design. $40-45.

97

Red geraniums in large blue pots on brick wall background. $40-45.

Tennis, golf bags, archery, sailing, and horses. Orange, blue, green on white. Old time sports. $40-50.

Very dark navy and yellow on dark gray with ladies and village of huts, carriages, and houses. $45-65.

98

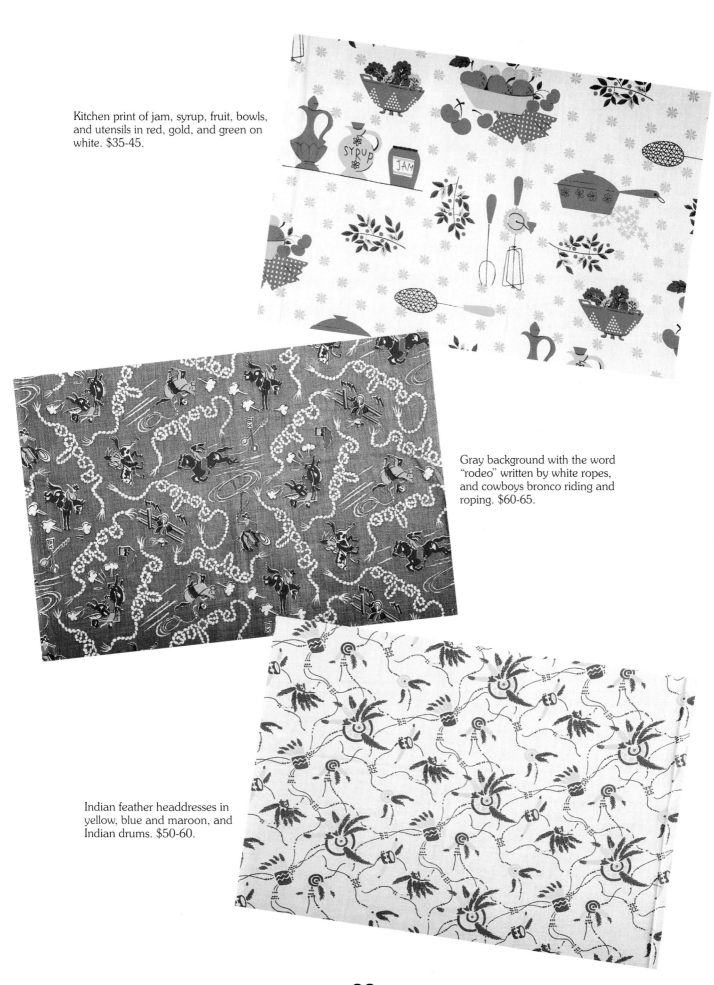

Kitchen print of jam, syrup, fruit, bowls, and utensils in red, gold, and green on white. $35-45.

Gray background with the word "rodeo" written by white ropes, and cowboys bronco riding and roping. $60-65.

Indian feather headdresses in yellow, blue and maroon, and Indian drums. $50-60.

Jack and Jill went up the hill, the water well, and Jack getting a spanking, all in pink, blue and white. $55-65.

Seed packages and gardening tools, wheel barrels, and sprinkling can. $35-45.

Feathers with letters on them and colleges such as Yale and Columbia. $35-45.

Yellow, blue, white, and pink waves with black shells and various species of fish. $45-60.

100

Small yellow, white and turquoise flare skirts on ballerinas with pink background. $40-55.

Pink with maroon and white pictures of palm trees, donkey carts, and men wearing sombreros. $40-55.

Fancy barn turrets on yellow. $30-40.

Blue, yellow and orange southwest print with siesta men, cactus, maracas, sombreros, guitars, and flowers. $45-60.

Beach scenes with palm trees in frames, lavender and turquoise décor. $35-50.

Large blue and white papale pods on pink $30-40.

Blue and white wooden buckets full of blue flowers on pink. $35-45.

Orange dressed ladies with fruit bowls on heads and brown houses and turquoise and yellow palms. $45-50.

Blue and maroon art deco print with suns, stars, and planets. $30-45

Blue hats, maroon birds, and large butterflies on a patio scene. $30-45.

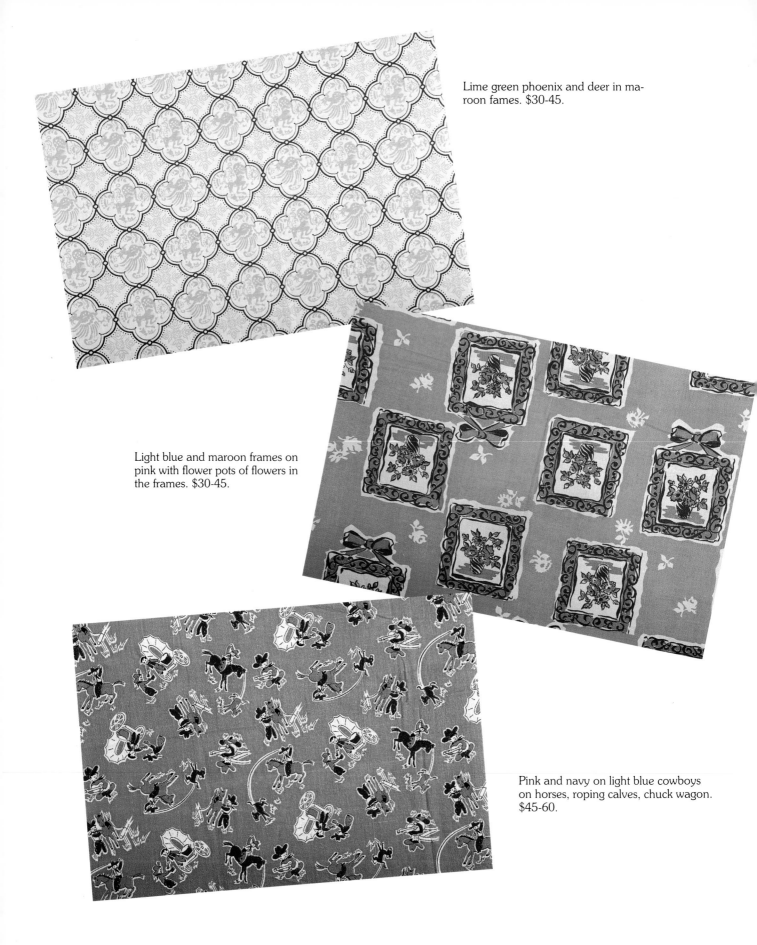

Lime green phoenix and deer in maroon fames. $30-45.

Light blue and maroon frames on pink with flower pots of flowers in the frames. $30-45.

Pink and navy on light blue cowboys on horses, roping calves, chuck wagon. $45-60.

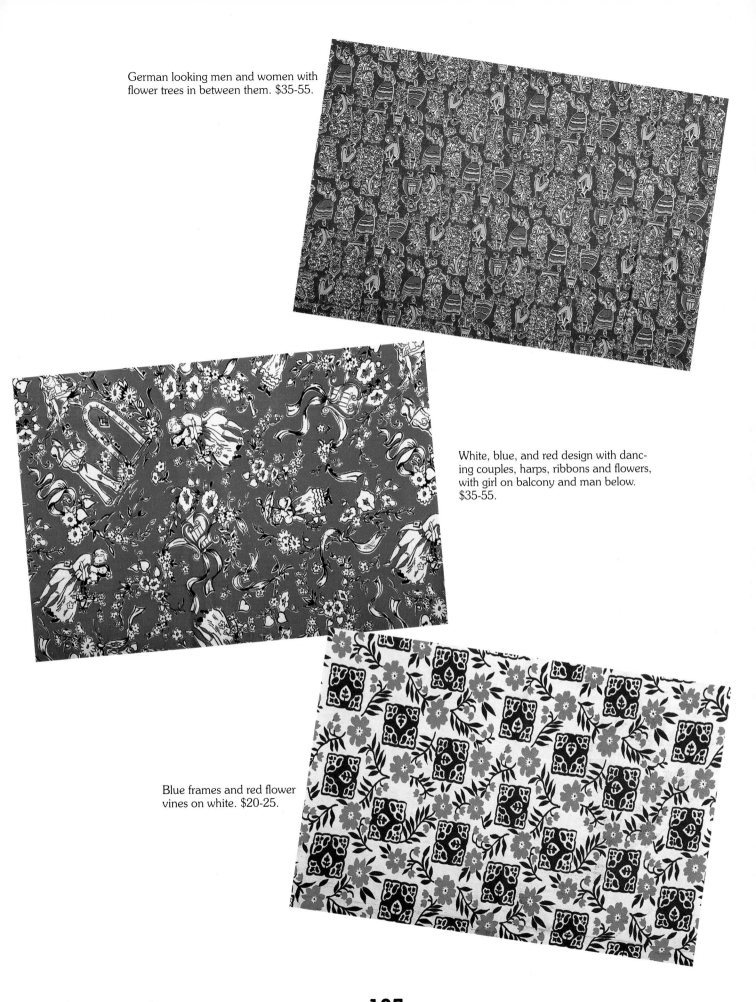

German looking men and women with flower trees in between them. $35-55.

White, blue, and red design with dancing couples, harps, ribbons and flowers, with girl on balcony and man below. $35-55.

Blue frames and red flower vines on white. $20-25.

Geometric Print Chicken Feed Sacks

Any designs that you can think of come in these bags. Designers went wild on many.

Blue and gray diagonal plaid design. $25-30.

Lime green polka dots on white. $20-35.

Blue and lavender stripes with red gingham squares. $35-40.

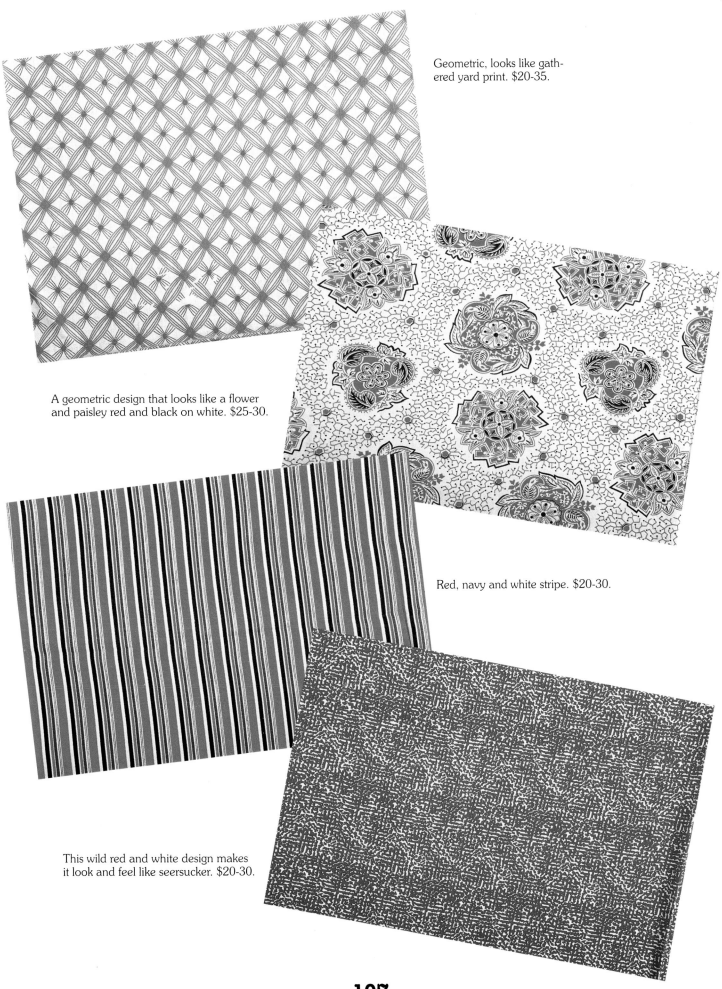

Geometric, looks like gathered yard print. $20-35.

A geometric design that looks like a flower and paisley red and black on white. $25-30.

Red, navy and white stripe. $20-30.

This wild red and white design makes it look and feel like seersucker. $20-30.

Someone called this Mickey Mouse ears. Really it is just a geometric design. $35-40.

Varying shades of pink ovals on white. $35-55.

Jagged scrolls on a pretty blue/ green. $25-35.

White cornucopia on gray. $20-25.

108

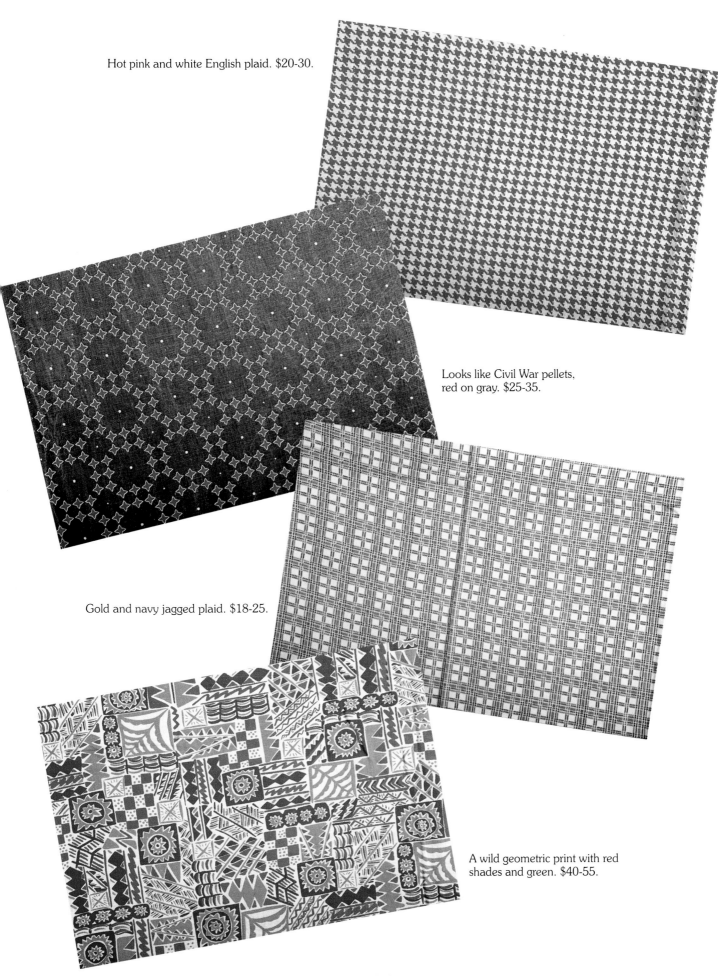

Hot pink and white English plaid. $20-30.

Looks like Civil War pellets, red on gray. $25-35.

Gold and navy jagged plaid. $18-25.

A wild geometric print with red shades and green. $40-55.

Brown, yellow and white large diamonds. $20-30

Green and white plaid. $45-55.

Red and white plaid. These really are feed sacks not tablecloth fabric. $45-55.

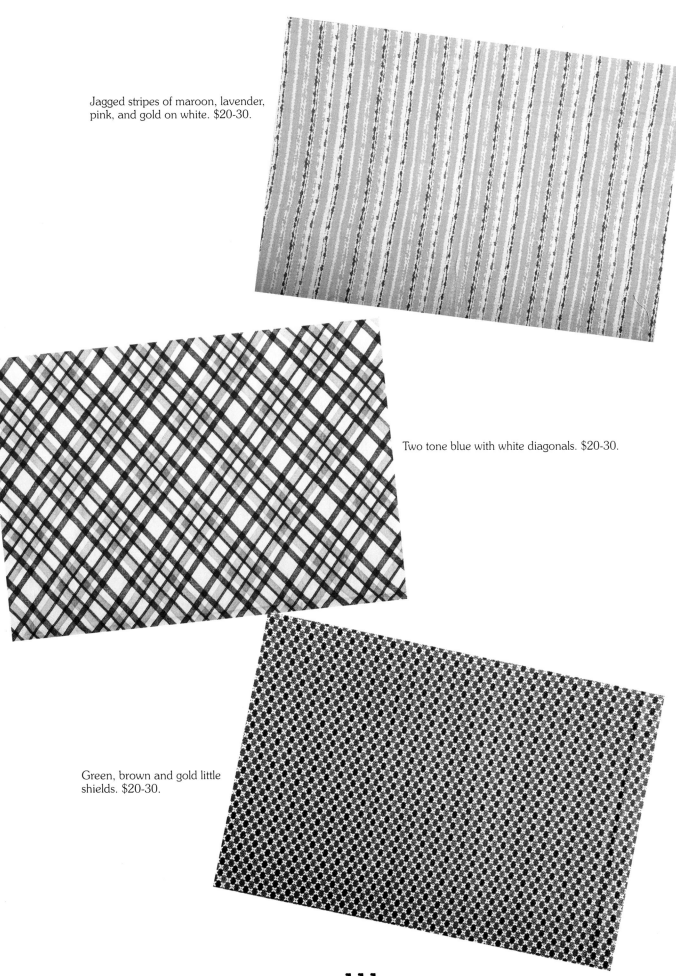

Jagged stripes of maroon, lavender, pink, and gold on white. $20-30.

Two tone blue with white diagonals. $20-30.

Green, brown and gold little shields. $20-30.

Green and white diagonals
with black X's. $20-30.

Geometric brown, tan, and white circles. $20-30.

Olive green and gold ovals,
and squares. $20-30.

Blue, red and white diagonal gingham. $18-25.

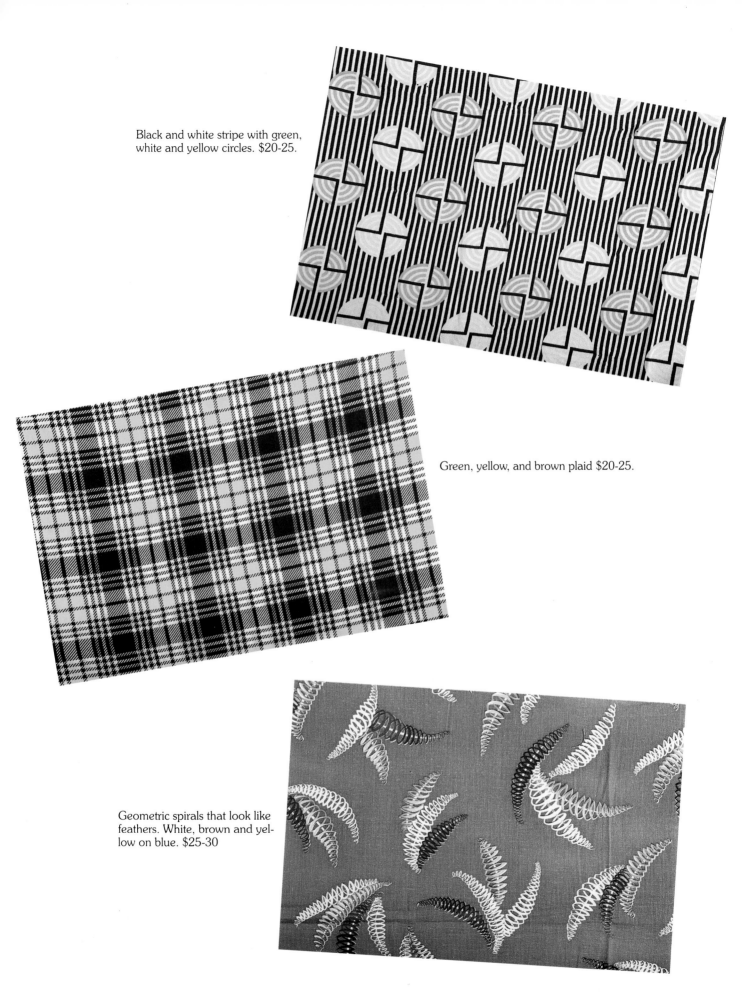

Black and white stripe with green, white and yellow circles. $20-25.

Green, yellow, and brown plaid $20-25.

Geometric spirals that look like feathers. White, brown and yellow on blue. $25-30

113

Pillowcase Chicken Feed Sacks

Many border prints were specifically meant to be pillowcases, while others you can tell were not really meant for pillows. These may be the least expensive but don't leave them out of your collecting if you want the full history.

Pillowcase; green and pink flowers scattered overall with pink and green butterflies border and scalloped green end. $30-45.

Pillowcase; pink floral bouquets with pink rose design solid end. $30-45.

Pillowcase; green roses design with solid green end. $30-40.

Pillowcase; blue and pink flowers scattered overall with pink and blue butterflies border and solid pink end. $30-45.

Pillowcase; tiny blue and yellow flowers above blue solid end. $18-25.

Pillowcase with overall rose bouquets design and blue scalloped edge with pink ribbon. $25-35

Pillowcase; tiny pink and yellow flowers above pink solid end. $18-25.

Pillowcase; turquoise blue rose bundles on pink ribbon with turquoise blue solid end. $18-25.

Pillowcase; yellow flower carts loaded with pink and yellow flowers and yellow solid end. $45-55.

Pillowcase; beautiful yellow roses in blue baskets above yellow scalloped end. $30-40.

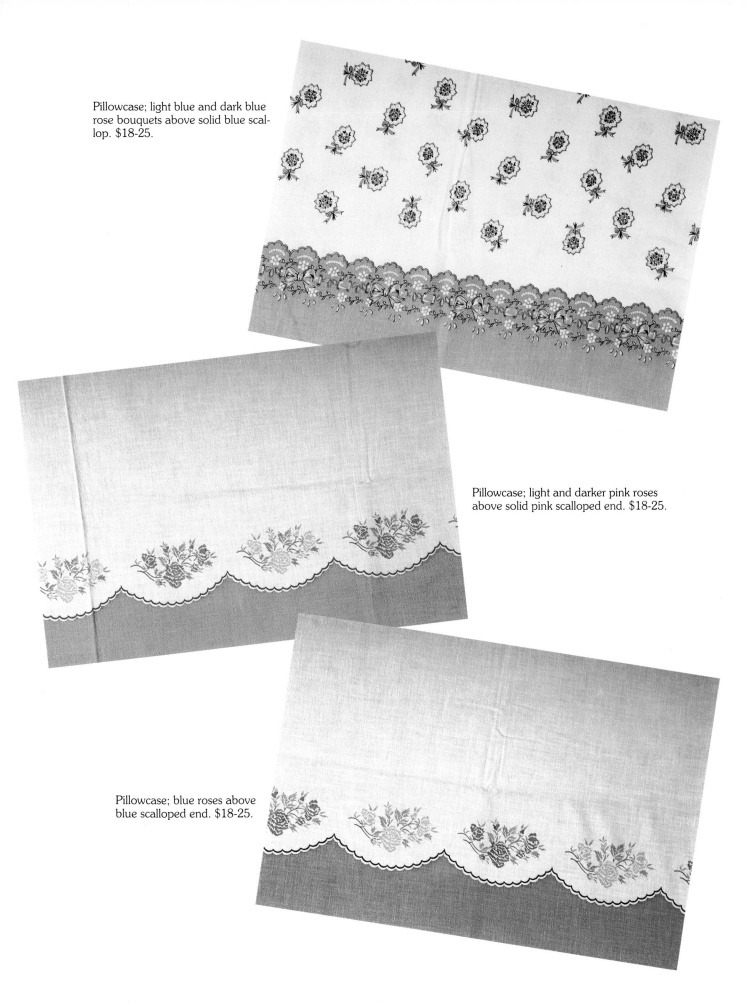

Pillowcase; light blue and dark blue rose bouquets above solid blue scallop. $18-25.

Pillowcase; light and darker pink roses above solid pink scalloped end. $18-25.

Pillowcase; blue roses above blue scalloped end. $18-25.

Florals Or Flowers Chicken Feed Sacks

Probably the easiest to find are the floral sacks. Sorry, I do not know the names of flowers better or I could have described them in more detail.

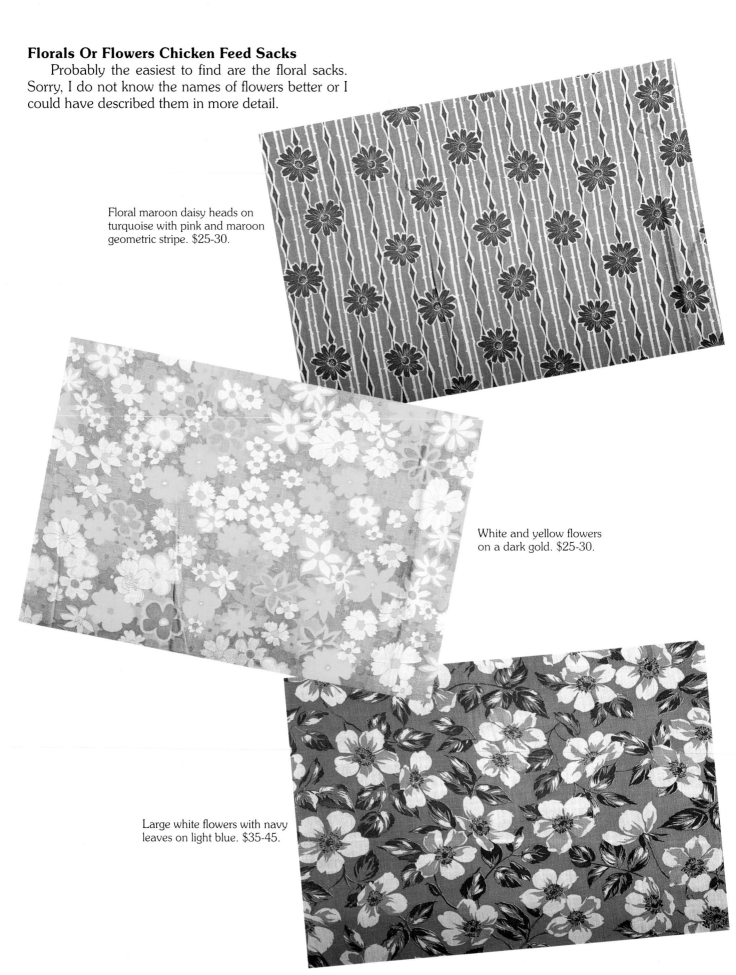

Floral maroon daisy heads on turquoise with pink and maroon geometric stripe. $25-30.

White and yellow flowers on a dark gold. $25-30.

Large white flowers with navy leaves on light blue. $35-45.

Dark pink petals outlined in gray between green and gray floral bouquets. $35-40.

Tiny yellow flowers and green leaves inside blue oval frames. $25-30.

Bright daisies on pink and white gingham. $35-45.

Blue turquoise and green pods with green leaves and vines on white. $25-35.

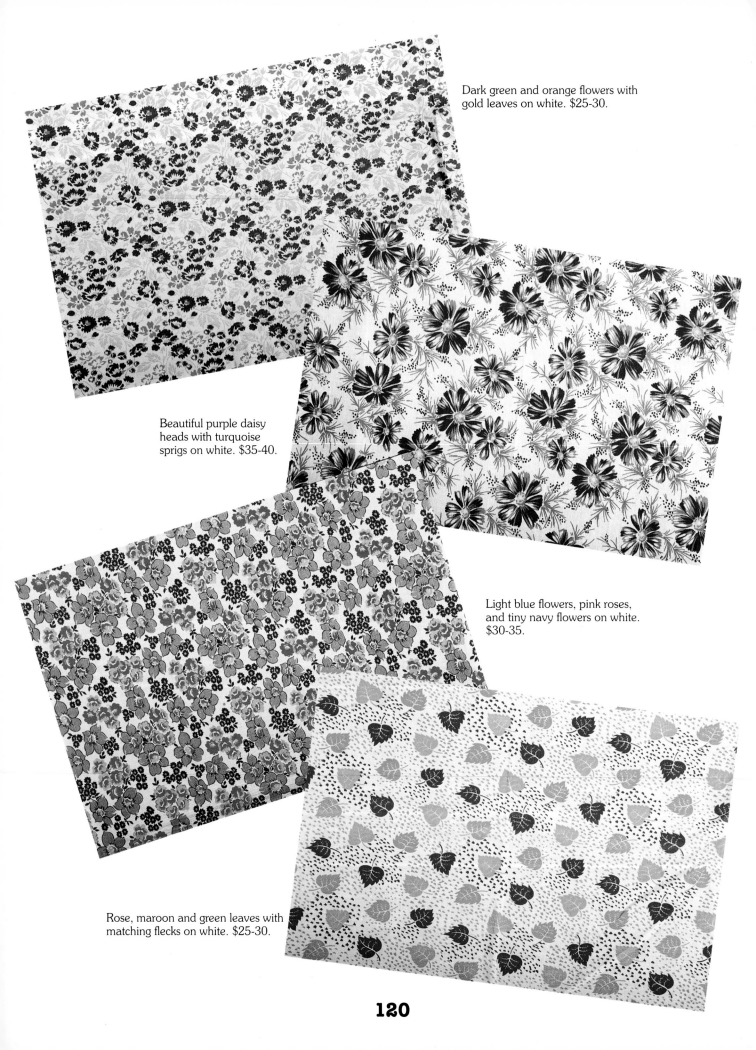

Dark green and orange flowers with gold leaves on white. $25-30.

Beautiful purple daisy heads with turquoise sprigs on white. $35-40.

Light blue flowers, pink roses, and tiny navy flowers on white. $30-35.

Rose, maroon and green leaves with matching flecks on white. $25-30.

Yellow and orange flowers with brown leaves on white. $25-30.

White palms and ferns on red. $25-35.

Lavender tulips with small pink and white flowers. $35-40.

Red and pink flowers and yellow flowers on green kite shapes and circles between. $30-35.

Delicate blue blossoms with yellow leaf bouquets and white waves between. $30-35.

Large yellow daisies on blue and pink basket weave. $25-30.

Turquoise daisies on gray and yellow geometric circles. $25-30.

Red tulips on gray striking print. $30-40.

Ball flowers in rose and turquoise on lavender. $25-30.

Bright daisies with green centers on navy blocks with dots. $25-30.

Overall orange and lavender flowers on white. $25-30.

Lavender palm tops on white. $25-30.

Green and white leaves surrounded geometric navy designs. $25-30.

Overall dainty pink flowers. $20-25.

Purple feather flowers and turquoise straw flowers on white. $35-40.

Tiny blue, orange, and green flowers between lines of diamonds. $30-35.

Large purple daisies on white. $35-45.

Large red flowers with blue centers with smaller purple and red outlined flowers on vines on white. $30-35.

Orange flower bunches with blue leaves on white. $30-35.

Blue rose bouquets on white. $30-35.

Stacked flowers and geometric designs. Straw flowers in blue and white with blue lines, on white. $30-35.

Large brown leaves on tan and green diagonal plaid. $30-35.

Floral large yellow poppies and red straw flowers on white. $30-35.

Red and pink floral bouquets tied with blue and pink bows on blue and white waving diamond plaid. $30-35.

Large yellow roses with sprawling turquoise leaves on white. $35-40.

Lavender and red bouquets on white. $30-35.

Purple flower heads on long green stems on white. $35-40.

Solid lavender with pink roses and tiny white petal flowers. $30-35.

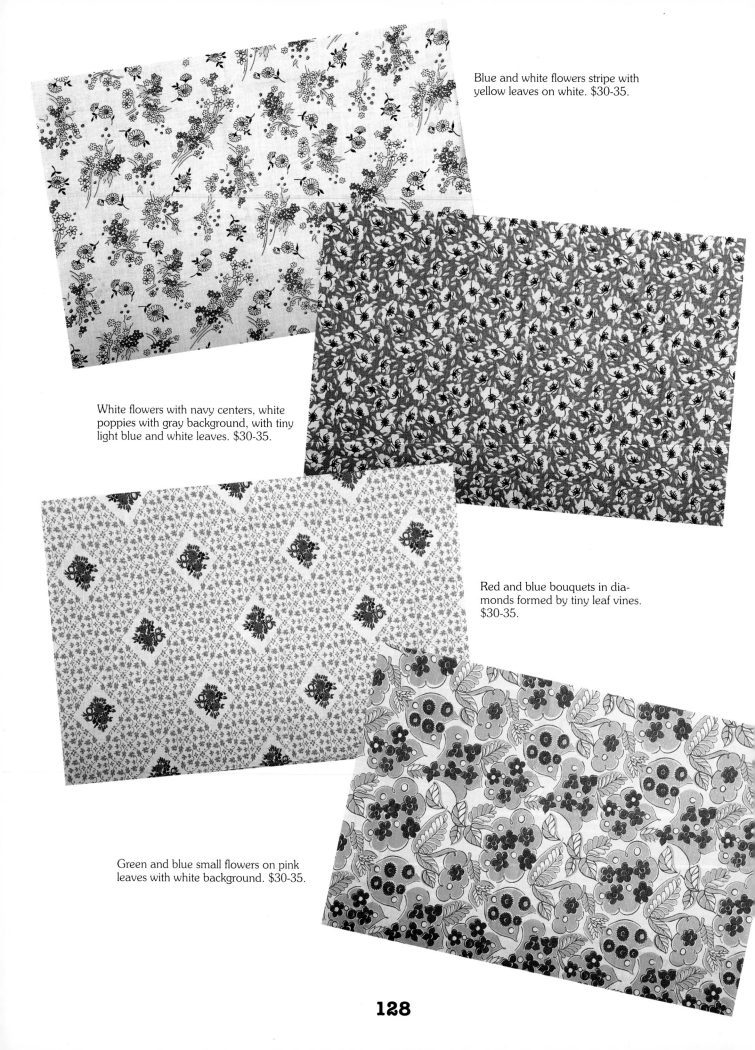

Blue and white flowers stripe with yellow leaves on white. $30-35.

White flowers with navy centers, white poppies with gray background, with tiny light blue and white leaves. $30-35.

Red and blue bouquets in diamonds formed by tiny leaf vines. $30-35.

Green and blue small flowers on pink leaves with white background. $30-35.

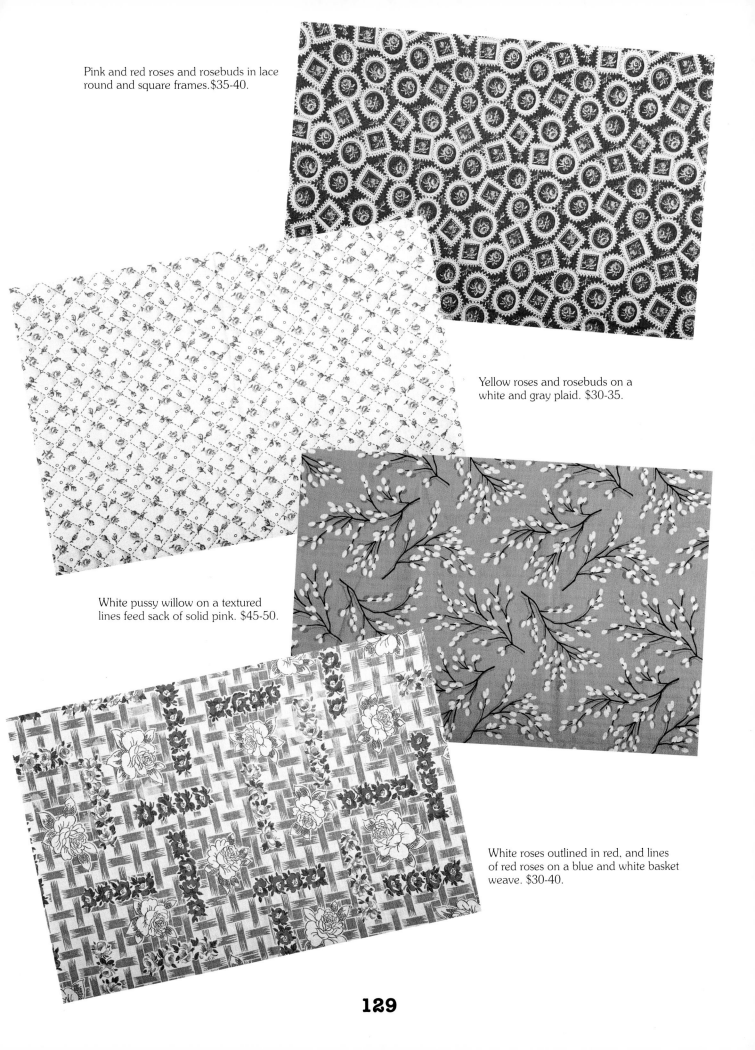

Pink and red roses and rosebuds in lace round and square frames. $35-40.

Yellow roses and rosebuds on a white and gray plaid. $30-35.

White pussy willow on a textured lines feed sack of solid pink. $45-50.

White roses outlined in red, and lines of red roses on a blue and white basket weave. $30-40.

129

Blue petal flowers sprinkled on pink petals on white. $30-35.

Red and brown flowers in white frames on turquoise. $35-40.

Blue tulips with green petal flowers in bouquets with pink dotted plaid. $30-35.

Two tone blue tulips and petal flowers, solid and bright on white. $35-40.

Turquoise and green flowers, large on white with brown sprigs and leaves. $30-35.

White flowers with purple and red centers, small and purple and white flowers all on solid red. $30-35.

Turquoise and white daisies with brown leaves. $30-35.

Navy and white half flowers with green centers on a scrolled geometric solid pink and pink dots on white. $30-35.

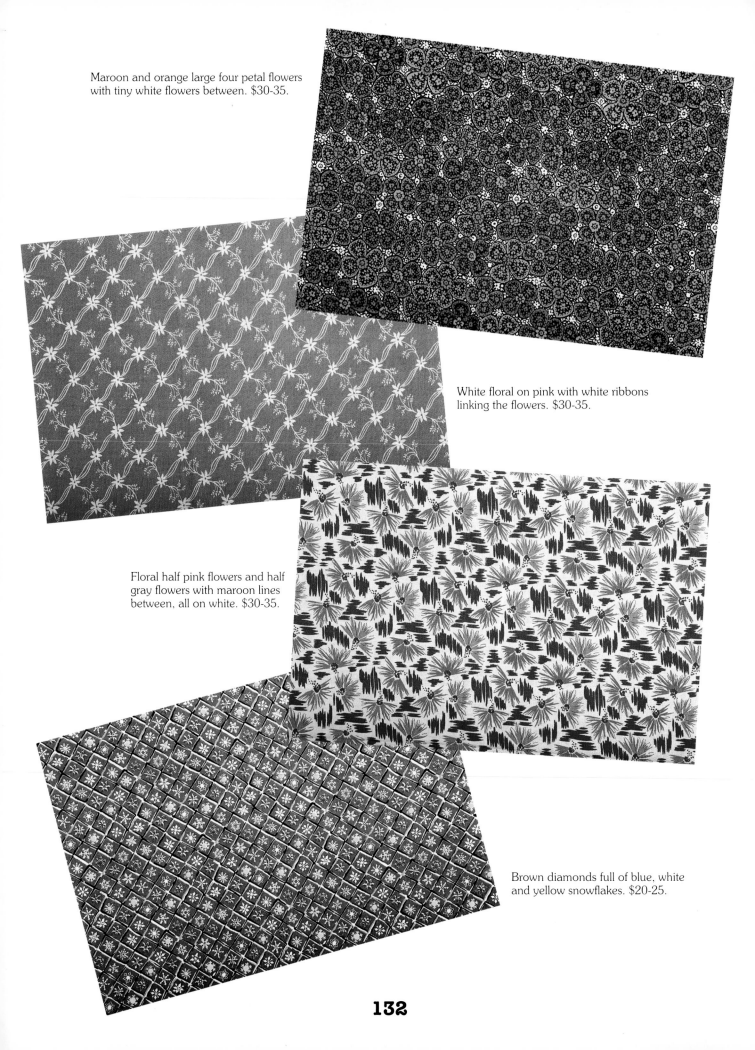

Maroon and orange large four petal flowers with tiny white flowers between. $30-35.

White floral on pink with white ribbons linking the flowers. $30-35.

Floral half pink flowers and half gray flowers with maroon lines between, all on white. $30-35.

Brown diamonds full of blue, white and yellow snowflakes. $20-25.

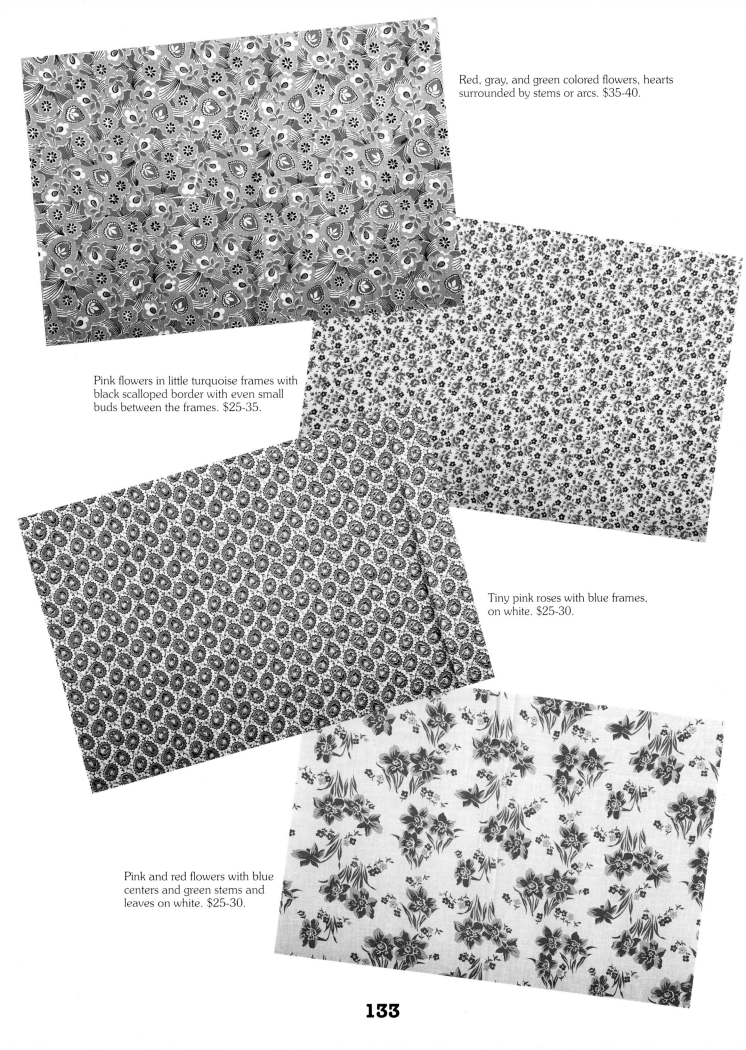

Red, gray, and green colored flowers, hearts surrounded by stems or arcs. $35-40.

Pink flowers in little turquoise frames with black scalloped border with even small buds between the frames. $25-35.

Tiny pink roses with blue frames, on white. $25-30.

Pink and red flowers with blue centers and green stems and leaves on white. $25-30.

133

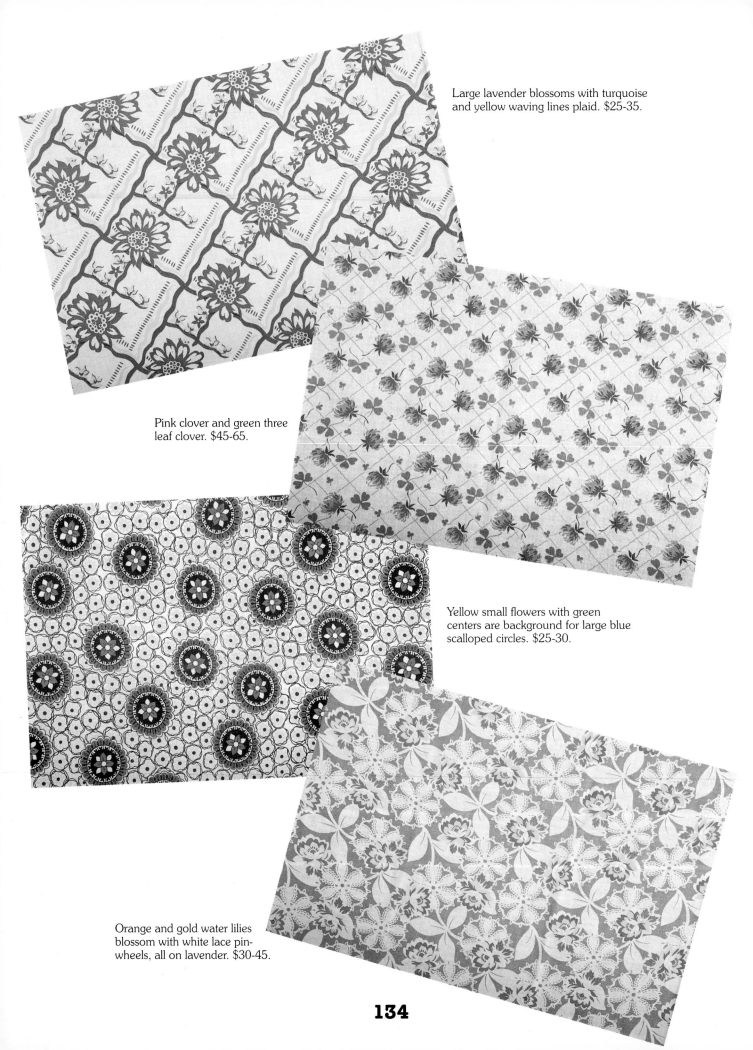

Large lavender blossoms with turquoise and yellow waving lines plaid. $25-35.

Pink clover and green three leaf clover. $45-65.

Yellow small flowers with green centers are background for large blue scalloped circles. $25-30.

Orange and gold water lilies blossom with white lace pin-wheels, all on lavender. $30-45.

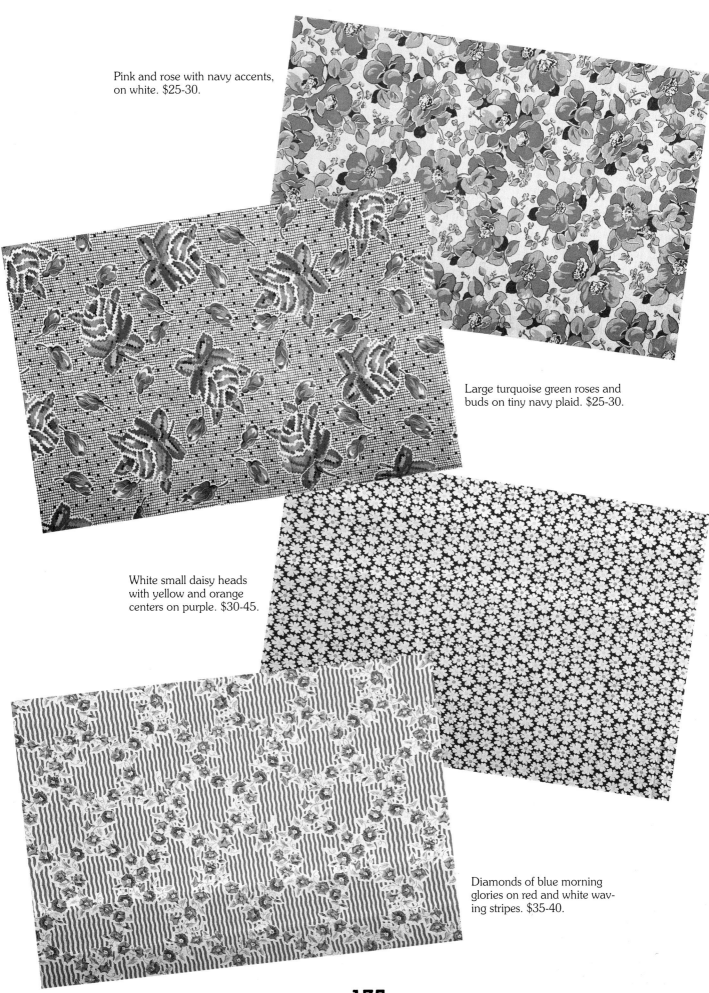

Pink and rose with navy accents, on white. $25-30.

Large turquoise green roses and buds on tiny navy plaid. $25-30.

White small daisy heads with yellow and orange centers on purple. $30-45.

Diamonds of blue morning glories on red and white waving stripes. $35-40.

Large yellow roses with navy leaves on white background with a rather gauzy texture. $30-40.

Purple, lime green, and white flowers in plaids. $25-35.

Three styles of bouquets in yellow, pink and maroon, all on blue. $30-40.

White flowers with green centers and wild geometric designs in red and yellow. $35-45.

Red and white lilies on lime green. $25-35.

Red flowers with white centers, green and blue sprigs leaves, all on white. $30-35.

Gold, purple, and red flowers with palm-like sprays of tiny flowers. $25-40.

Blue and cream stripes outlined in scalloped red with tiny blue and yellow floral vines in between. $25-30.

Floral with textured pattern and bright yellow and red flowers on blue. $45-50.

White frames of blue roses with white vines in between on blue. $25-40.

Bright bold red leaves with turquoise, brown and lime green flowers. $35-45.

Tiny pink groups with blue floral vines with green leaves on cream. $25-30.

138

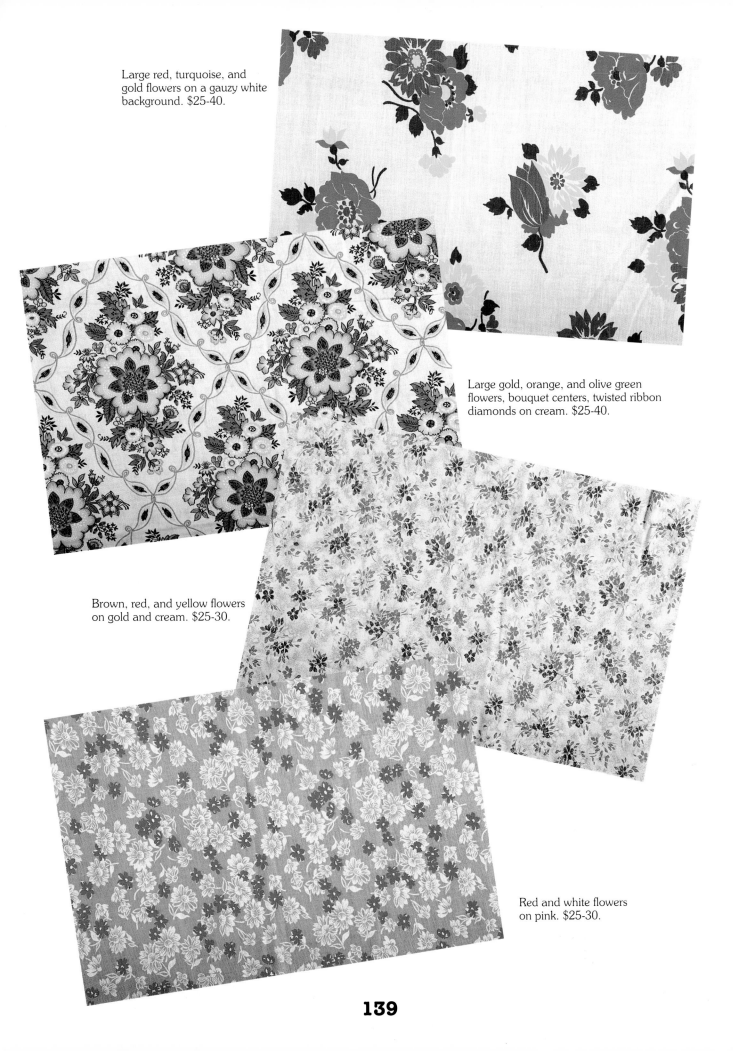

Large red, turquoise, and gold flowers on a gauzy white background. $25-40.

Large gold, orange, and olive green flowers, bouquet centers, twisted ribbon diamonds on cream. $25-40.

Brown, red, and yellow flowers on gold and cream. $25-30.

Red and white flowers on pink. $25-30.

Stripes of three red roses with green leaves on solid yellow. $25-30.

Gray and white roses on blue. $25-30.

Gray and white blossoms outlined with maroon, all on solid blue. $25-30.

Large white and blue flowers with red centers. $30-45.

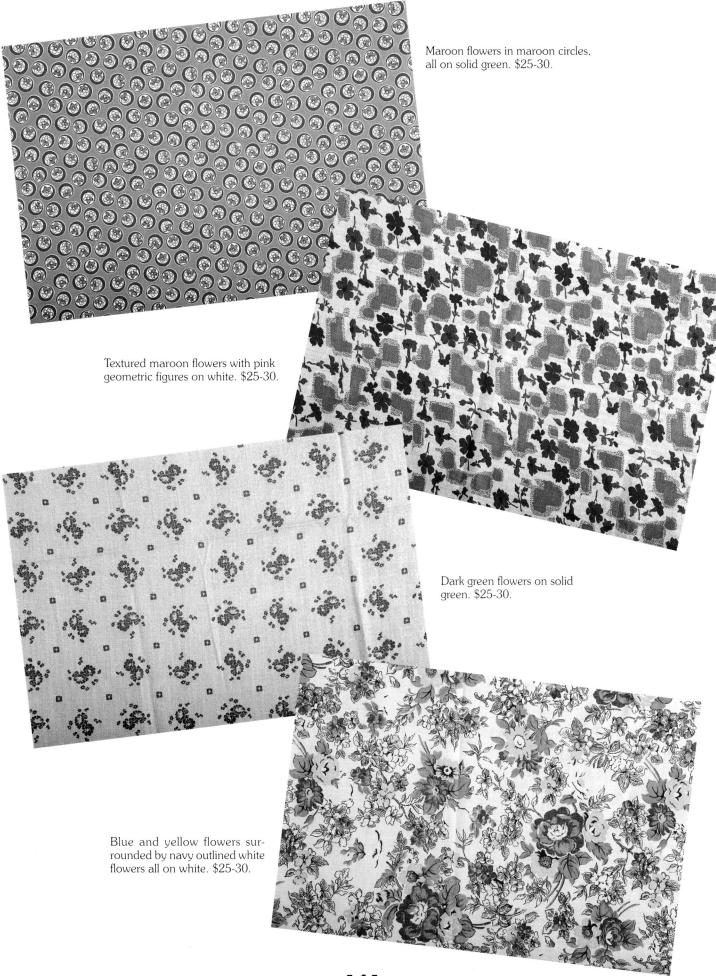

Maroon flowers in maroon circles, all on solid green. $25-30.

Textured maroon flowers with pink geometric figures on white. $25-30.

Dark green flowers on solid green. $25-30.

Blue and yellow flowers surrounded by navy outlined white flowers all on white. $25-30.

141

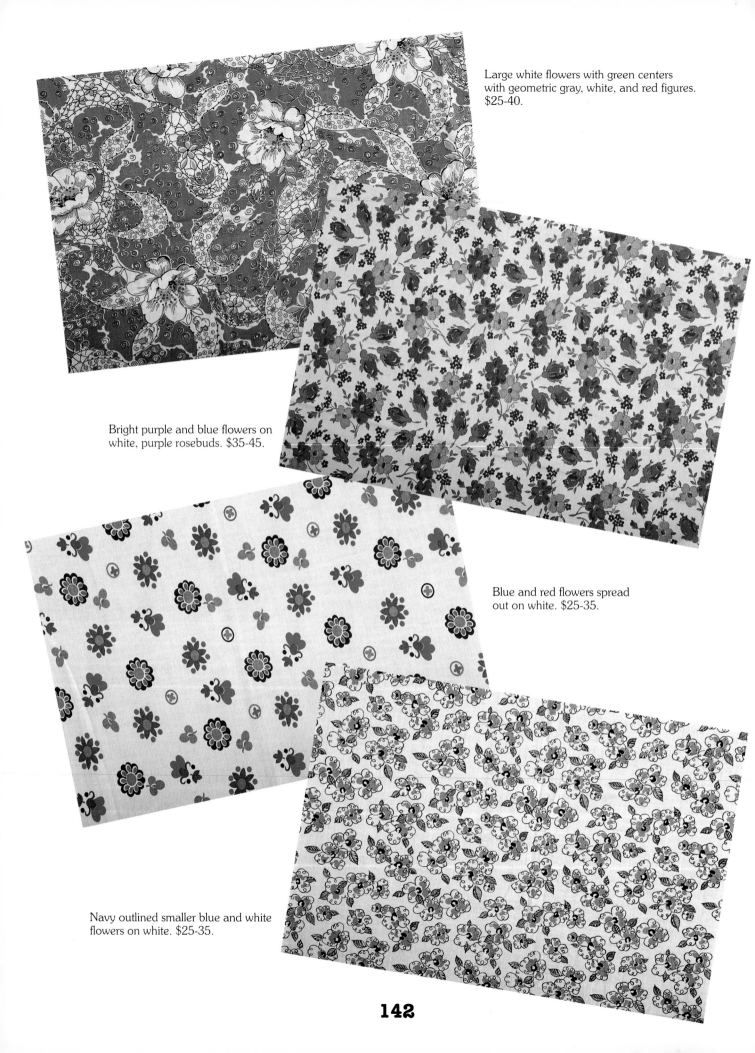

Large white flowers with green centers with geometric gray, white, and red figures. $25-40.

Bright purple and blue flowers on white, purple rosebuds. $35-45.

Blue and red flowers spread out on white. $25-35.

Navy outlined smaller blue and white flowers on white. $25-35.

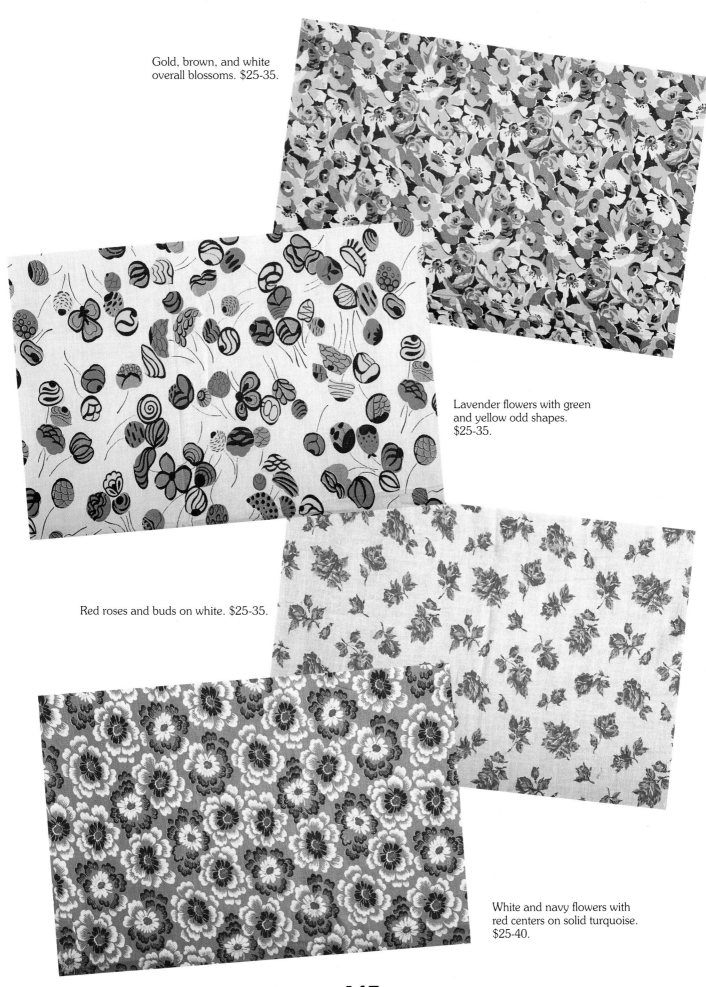

Gold, brown, and white overall blossoms. $25-35.

Lavender flowers with green and yellow odd shapes. $25-35.

Red roses and buds on white. $25-35.

White and navy flowers with red centers on solid turquoise. $25-40.

Yellow flowers with blue leaves on solid maroon. $25-35.

Purple and white flowers on white with solid purple flowers. $25-35.

Red and white flowers on white. $25-35.

Yellow, gray, and blue overall flowers. $25-35.

Blue and white tulips with pink leaves on maroon. $35-40.

Maroon outlined pink flowers and maroon and green leaves on white. $35-40.

Purple outlined gray flowers with purple leaves on white. $25-40.

Red rosebuds with tiny blue flowers on white. $25-35.

145

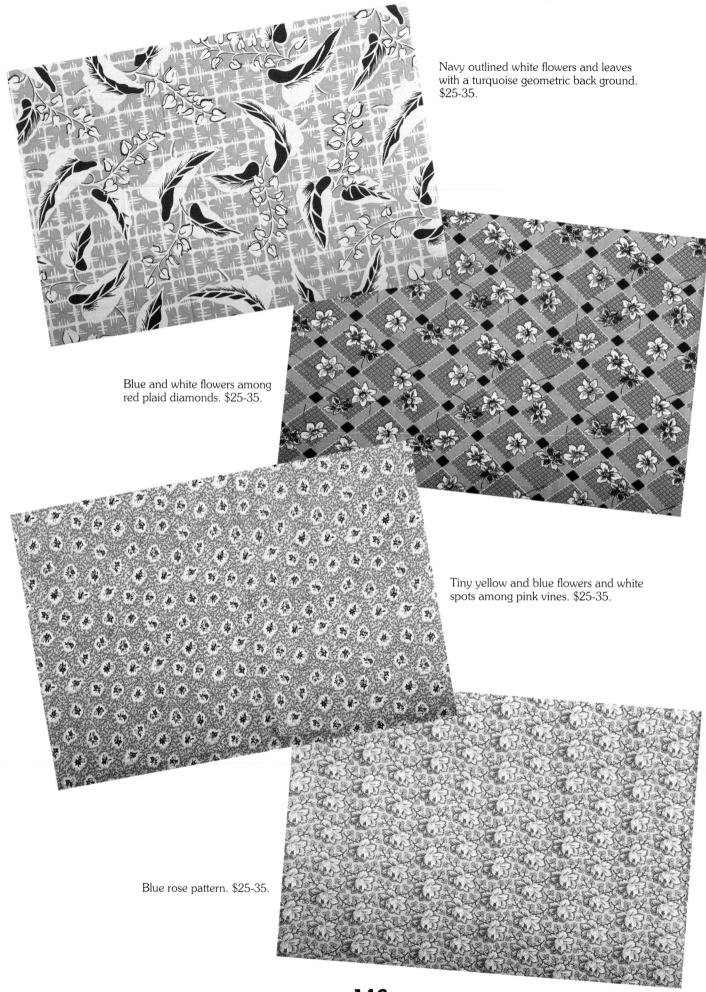

Navy outlined white flowers and leaves with a turquoise geometric back ground. $25-35.

Blue and white flowers among red plaid diamonds. $25-35.

Tiny yellow and blue flowers and white spots among pink vines. $25-35.

Blue rose pattern. $25-35.

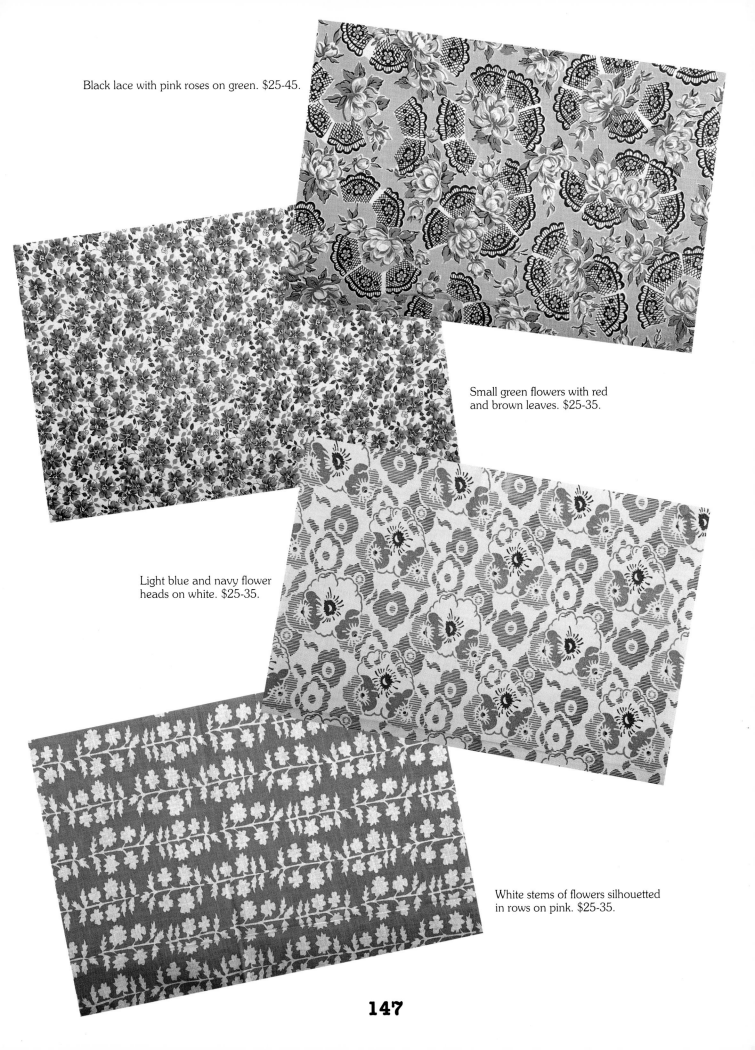

Black lace with pink roses on green. $25-45.

Small green flowers with red and brown leaves. $25-35.

Light blue and navy flower heads on white. $25-35.

White stems of flowers silhouetted in rows on pink. $25-35.

Border Prints Chicken Feed Sacks

This is a grouping of feed sacks with border prints that would be better off as curtains or something other than just pillowcases. Some could be both.

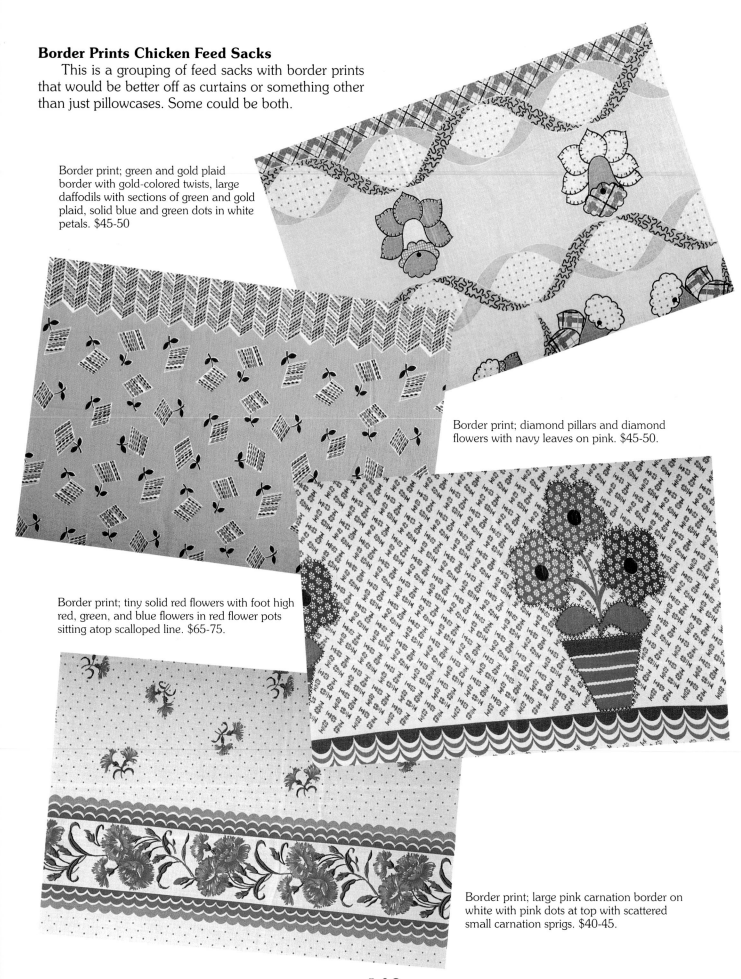

Border print; green and gold plaid border with gold-colored twists, large daffodils with sections of green and gold plaid, solid blue and green dots in white petals. $45-50

Border print; diamond pillars and diamond flowers with navy leaves on pink. $45-50.

Border print; tiny solid red flowers with foot high red, green, and blue flowers in red flower pots sitting atop scalloped line. $65-75.

Border print; large pink carnation border on white with pink dots at top with scattered small carnation sprigs. $40-45.

148

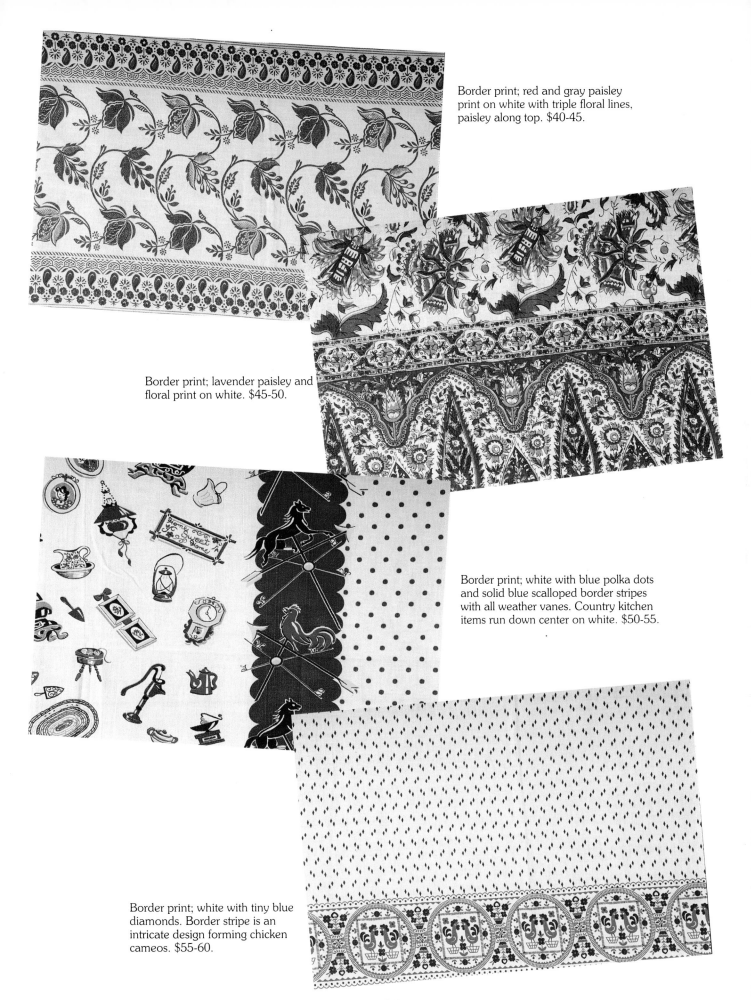

Border print; red and gray paisley print on white with triple floral lines, paisley along top. $40-45.

Border print; lavender paisley and floral print on white. $45-50.

Border print; white with blue polka dots and solid blue scalloped border stripes with all weather vanes. Country kitchen items run down center on white. $50-55.

Border print; white with tiny blue diamonds. Border stripe is an intricate design forming chicken cameos. $55-60.

Border print; green with scattered dots and moving up to decorated fruit designs and butterflies with yellow dots and navy circles on top of each other, on white. $45-50.

Border print; double rows on both sides and kitchen dishes down center with prominent red and green décor. $45-50.

Border print; white with brown polka dots and brown men. Lime colored ladies and trees. Men have top hats, waistcoats, and canes, and are holding flower bouquets. Ladies are shy with bonnets, scalloped designs on skirts, and closed parasols. $35-45.

Border print; very special red and blue hens with nests behind chicken wire. Two border rows on each side of white with blue dots. $75-80.

Border print; Dutch people spinning yarn, playing violin, gathering the yarn, churning butter, and carrying spindles. Rows of green and blue rickrack design. $65-70.

Border print; lavender daffodils, green leaves very thick at bottom, and becoming gradually thinner at top. $55-65.

Border print; blue and green border at bottom with dots on white at top. Bottom border comes up to vase full of red and green flowers. $45-50.

Border print; white with green polka dots with a dotted fruit with bright green and red leaves. This is a double border with the top line repeating at the other end also. $40-45.

151

Border print; bright pink on white petals on solid pink and large geometric design on white. $35-40.

Border print; two rows of gray ruffles with white dots on bright pink and white yarn diamonds. $30-35.

Border print; large white dotted red apples with tiny red apples on white. $40-45.

Border print; large roses and green leaves. $40-45.

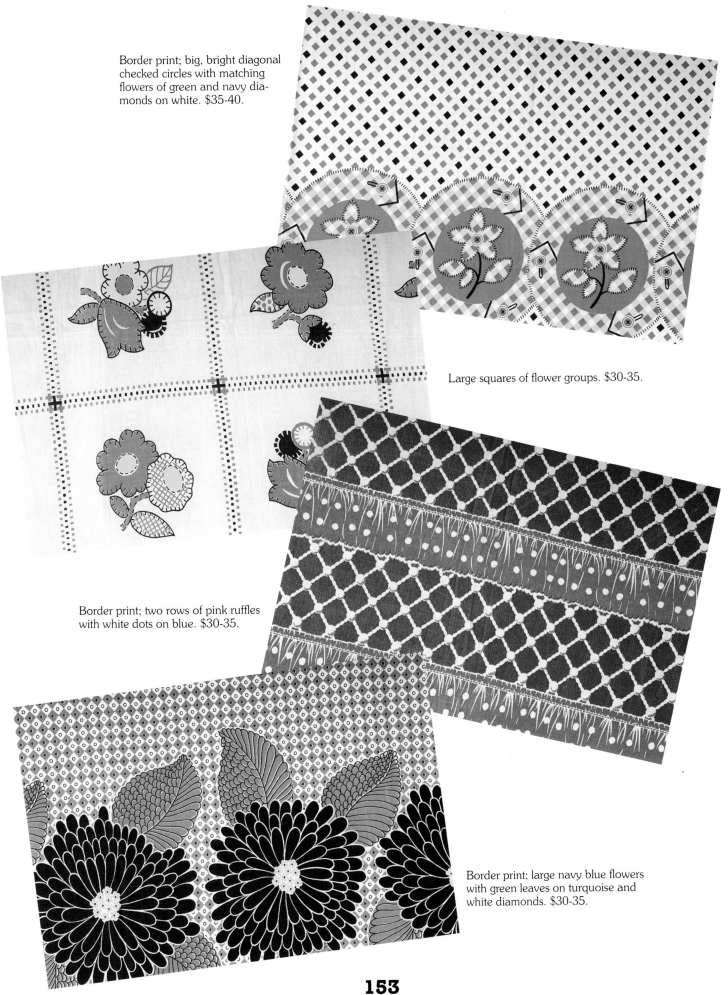

Border print; big, bright diagonal checked circles with matching flowers of green and navy diamonds on white. $35-40.

Large squares of flower groups. $30-35.

Border print; two rows of pink ruffles with white dots on blue. $30-35.

Border print; large navy blue flowers with green leaves on turquoise and white diamonds. $30-35.

Fruit Prints Chicken Feed Sacks

In no way were the designers confined to actual colors when it came to making fruit prints. Who knew fruit could be plaid?

Tiny apples with red flowers on white. $35-40.

White grapes and leaves on solid orange. $35-40.

Purple plums on white. $30-35.

154

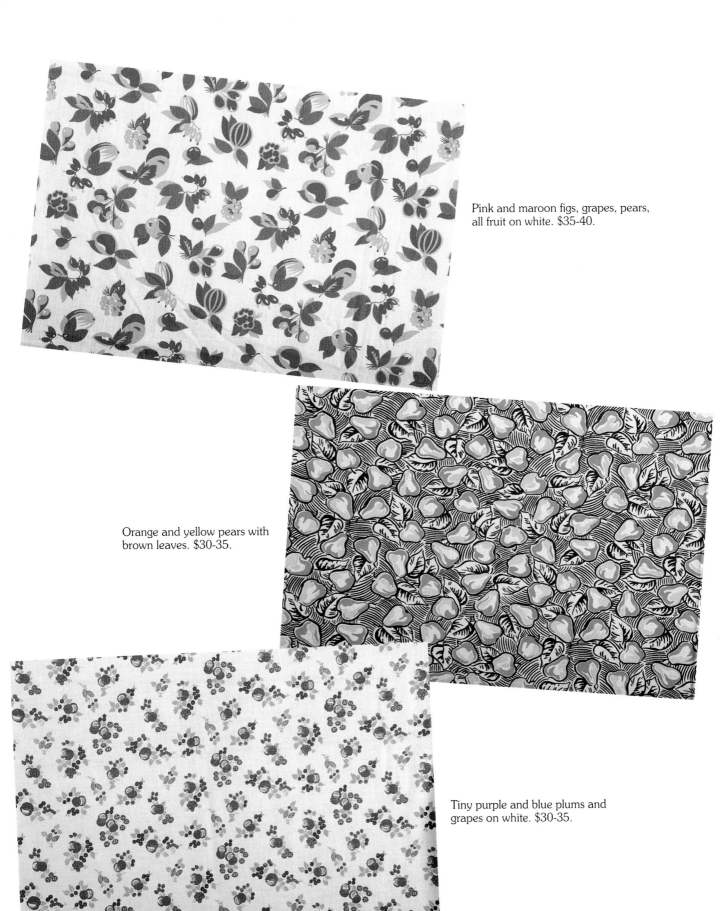

Pink and maroon figs, grapes, pears,
all fruit on white. $35-40.

Orange and yellow pears with
brown leaves. $30-35.

Tiny purple and blue plums and
grapes on white. $30-35.

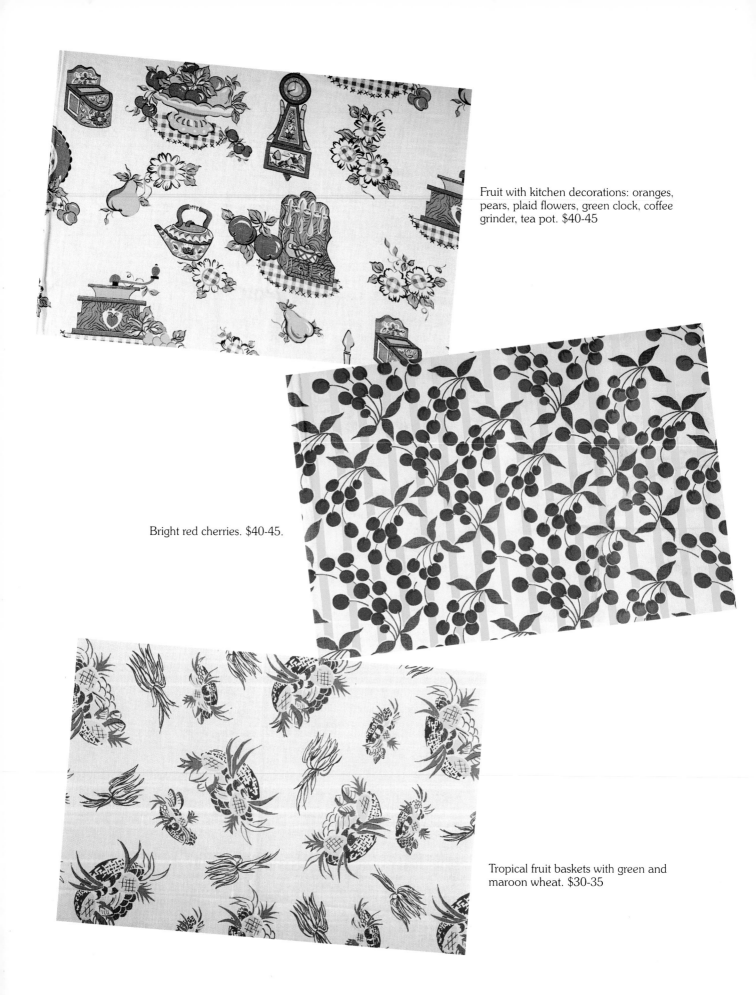

Fruit with kitchen decorations: oranges, pears, plaid flowers, green clock, coffee grinder, tea pot. $40-45

Bright red cherries. $40-45.

Tropical fruit baskets with green and maroon wheat. $30-35

Red strawberries and green leaves on white. $50-60.

Plaid fruit: oranges, peaches, yellow pineapple, brown apples. $30-35.

Plaid fruit: oranges, brown apples, gold pears, brown leaves on white. $40-45.

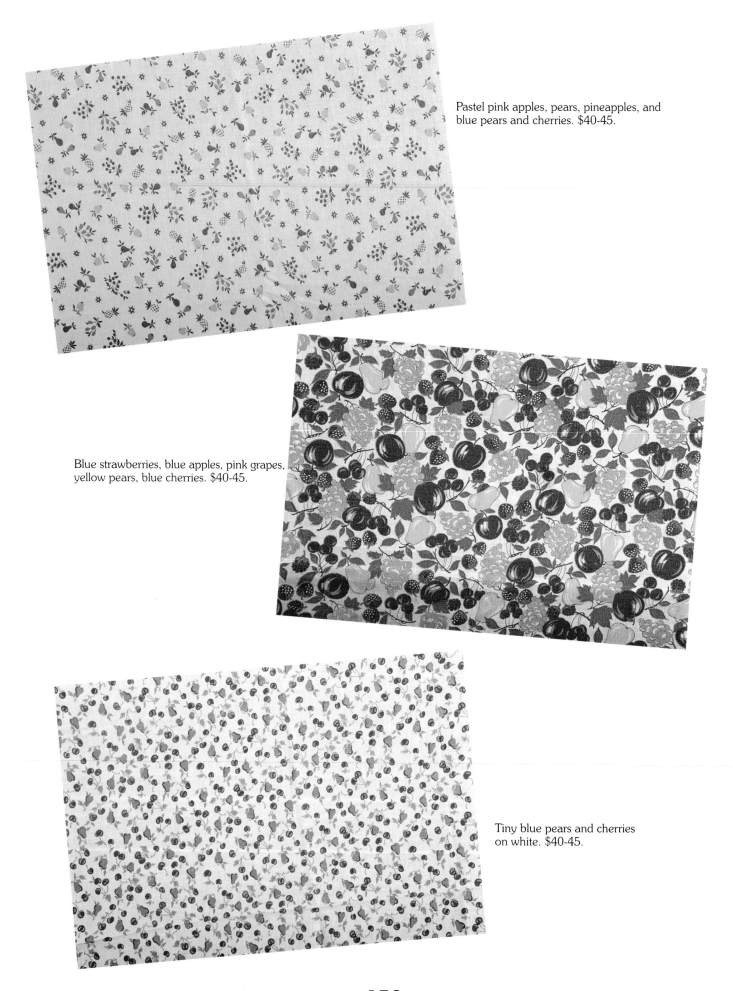

Pastel pink apples, pears, pineapples, and blue pears and cherries. $40-45.

Blue strawberries, blue apples, pink grapes, yellow pears, blue cherries. $40-45.

Tiny blue pears and cherries on white. $40-45.

Blue cherries on white. $35-40.

Fruit border print with red ribbons of purple grapes, green apples, and purple plums. $35-40.

Plain Solid Color Feed Sacks

Oddly, this is probably the hardest sack to find for your collection. If you are a purist you probably want whatever you are making to be made entirely of feed sacks. If a solid is required you may find this difficult. Some of the colors may be simply different shades due to the degree of fade present. However, solid colors probably exist as shades of yellow, green, red, and blue. 1950s colors, like in pottery, are rose, gray, lavender or purple, plus different shades of green, turquoise, and gold.

Solid prints; blue, gold, yellow. $60-75.

Solid prints; blue, gray, pink, green, gray. $60-75

Remember To Write

Please remember to write to me (via the publisher) so this fabulous journey can continue and we can learn more about cloth bags and feed sacks.